AF748987

"This groundbreaking book delves into Nepal's environmental challenges, providing valuable frameworks and lessons for global scholars, policymakers, and practitioners. Offering insights into the complex issues surrounding environmental justice issues in Nepal, the book discusses potential solutions for achieving just and equitable outcomes from environmental conservation."

Pema Gyamtsho, PhD,
Director General, ICIMOD, Nepal

"The book uncovers the untold miseries of environmental inequity and injustice faced by the rural and urban poor in Nepal due to soil erosion, habitat destruction, deforestation, hazardous use of pesticides and chemicals as well as ill-conceived and implemented development projects. While the publication is Nepal focused, the lessons learned can be extremely valuable to other countries as well."

Hon. Kaylan Shrestha,
Former Senior Justice Nepal Supreme Court

"Here is a vital contribution to global Environment Justice (EJ) scholarship that takes seriously the axes of caste, ethnicity, gender, and internal colonialism in the making of land, livelihood, and resource struggles. In thinking critically across Nepal's myriad political and liberation ecologies, this pathbreaking volume deepens our understanding of EJ both within and beyond the west."

Malini Ranganathan, *Associate Professor,*
American University, USA

"*Environmental Justice in Nepal* is a stunning and important contribution to global environmental justice scholarship. Grounded from the voices and standpoints of Nepalese activists and scholars, the collection not only addresses a wide range of topics (climate, land, health, conservation, development, land rights etc.), the volume collectively offers a new, unique and significant perspective on social dynamics, histories, and controversies of a vitally important place."

Julie Sze, *Professor, American Studies,*
UC Davis, USA

"This extraordinary volume features scholarship and activism that chart an inspiring course for environmental justice in Nepal. The contributors powerfully demonstrate how diverse Nepali communities boldly confront ecological and climate threats intertwined with patriarchy, environmental casteism, and internal colonialism to promote innovative pathways toward environmental quality and dignified livelihoods."

David Naguib Pellow, *Professor, Environmental Studies,*
UC Santa Barbara, USA

"We in the media used the acronym 'EJ' to mean 'Environmental Journalism'. Now I realise that it also stands for 'Environmental Justice'. This book is a must-read for us to understand that nature conservation and social justice are two sides of the same coin."

Kunda Dixit, *Author of* Dateline Earth: Journalism as If the Planet Mattered

"*Environmental Justice in Nepal* builds a transdisciplinary lens on environmental justice from uniquely Nepali standpoints that centre subaltern knowledge and experience. The chapters feature invaluable case studies that collectively establish Nepal as a crucial site of scholarly innovation for thinking through today's planetary environmental challenges. Indispensable reading for students, activists, planners and scholars."

Katherine Rankin, *Professor, Department of Geography and Program in Planning, University of Toronto, Canada*

Environmental Justice in Nepal

This edited volume provides a holistic compilation of the diverse range of emerging scholarship in critical environmental justice studies in Nepal.

This book brings together environmental justice scholarship set within a robust conceptual framework, focusing on a diversity of case studies from Nepal. Its locale-specific contextualization provides a unique analysis of the natural resource-based livelihoods common in the region, together with the health and well-being impacts of urban and industrial developments in its rapidly changing political, economic, social, and ecological environment. Centering contributions from Nepalese scholars and practitioners, this volume spans a wide range of topics, including the origins of environmental justice in Nepal, land and agriculture, conservation, infrastructure and development, Indigenous peoples, climate justice, and health equity. It reflects on the rise and development of social movements and public policy, discusses the further evolution of environmental justice, and highlights how the work of scholars, activists, and practitioners in the Nepalese context can enrich global conversations about social and environmental issues.

This book will appeal to scholars, researchers, students, and activists in environmental justice, sustainable development, South Asian, and Himalayan studies.

Jonathan K London is Professor in the Department of Human Ecology/Community and Regional Planning at the University of California, Davis, USA.

Jagannath Adhikari works as an independent researcher and teaches in Nepal and Australia.

Thomas Robertson is a historian and the former director of Fulbright Nepal/USEF.

Routledge Studies in Environmental Justice

This series is theoretically and geographically broad in scope, seeking to explore the emerging debates, controversies and practical solutions within Environmental Justice, from around the globe. It offers cutting-edge perspectives at both a local and global scale, engaging with topics such as climate justice, water governance, air pollution, waste management, environmental crime, and the various intersections of the field with related disciplines.

The *Routledge Studies in Environmental Justice* series welcomes submissions that combine strong academic theory with practical applications, and as such is relevant to a global readership of students, researchers, policy-makers, practitioners and activists.

For more information about this series, please visit: www.routledge.com/Routledge-Studies-in-Environmental-Justice/book-series/EJS

Environmental Justice in Nepal

Origins, Struggles, and Prospects

Edited by
Jonathan K London, Jagannath Adhikari, and Thomas Robertson

Routledge
Taylor & Francis Group
LONDON AND NEW YORK

earthscan
from Routledge

First published 2025
by Routledge
4 Park Square, Milton Park, Abingdon, Oxon OX14 4RN

and by Routledge
605 Third Avenue, New York, NY 10158

Routledge is an imprint of the Taylor & Francis Group, an informa business

© 2025 selection and editorial matter, Jonathan K London, Jagannath Adhikari, and Thomas Robertson; individual chapters, the contributors

The right of Jonathan K London, Jagannath Adhikari, and Thomas Robertson to be identified as the authors of the editorial material, and of the authors for their individual chapters, has been asserted in accordance with sections 77 and 78 of the Copyright, Designs and Patents Act 1988.

The Open Access version of this book, available at www.taylorfrancis.com, has been made available under a Creative Commons Attribution-Non Commercial-No Derivatives (CC-BY-NC-ND) 4.0 license.

This was funded by the University of California Davis.

Any third party material in this book is not included in the OA Creative Commons license, unless indicated otherwise in a credit line to the material. Please direct any permissions enquiries to the original rightsholder.

For Product Safety Concerns and Information please contact our EU representative GPSR@taylorandfrancis.com. Taylor & Francis Verlag GmbH, Kaufingerstraße 24, 80331 München, Germany.

Trademark notice: Product or corporate names may be trademarks or registered trademarks, and are used only for identification and explanation without intent to infringe.

British Library Cataloguing-in-Publication Data
A catalogue record for this book is available from the British Library

ISBN: 978-1-041-35155-9 (hbk)
ISBN: 978-1-032-44246-4 (pbk)
ISBN: 978-1-003-37117-5 (ebk)

DOI: 10.4324/9781003371175

Typeset in Times New Roman
by codeMantra

Contents

Figures

Tables

Contributors

Jagannath Adhikari works mainly as independent researcher and teaches in Nepal and Australia. He is interested in researching environmental justice, participatory natural resources management, food security, globalization and migration, and urban studies, and has published extensively on these themes.

Kishor Atreya is currently affiliated with the Institute of Forestry, Tribhuvan University, Nepal, where he serves as an associate professor of mountain environment and development. His research focuses on various areas, including the management of natural resources for livelihoods, traditional knowledge and socio-ecological systems, and the economic valuation of nature services and disservices.

Sonia Awale is Executive Editor of *Nepali Times* where she also serves as the health, science, and environment correspondent. She has extensively covered the climate crisis, disaster preparedness, development, and public health – looking at their political and economic interlinkages.

Srijana Baral is a researcher and practitioner of gender equality, social inclusion, and governance in forestry, climate change, and natural resource management. With almost 15 years of experience, she is currently a gender equality and social inclusion lead at RECOFTC, Thailand.

Narayan Belbase, an environmental public interest lawyer, has been working for over three decades toward facilitating and mainstreaming environmental governance and justice in Nepal's new Constitution. He is the author of over a dozen books on environmental policy, legislation, and gender justice.

Arjun Kumar B.K. is a writer, researcher, and advocate dedicated to uplifting marginalized communities in Nepal, particularly the Dalit community. He actively contributes to evidence-based policy advocacy at the Samata Foundation.

Sudikshya Bhandari is a trained architect and designer and a scholar in critical human geography with a focus on culture, housing, and sustainability. She is a doctoral candidate in the Geography Graduate Group at the University of California Davis.

Sunita Chaudhary is working as an ecosystem services specialist at the International Centre for Integrated Mountain Development (ICIMOD), Kathmandu, Nepal. Her area of expertise is interdisciplinary, focusing on human-nature interactions from the perspective of human geography.

Suresh Kumar Dhakal is an associate professor of Anthropology at Tribhuvan University. His current research interests include peasantry, land tenure/agrarian relations, bonded laborers, and environmental justice. He is also engaged in linking academic exercise to social issues through public debate.

Mandira Lamichhane Dhimal is Social Scientist and Policy Expert in Nepal. She has more than a decade experiences in social science research, interdisciplinary studies, and policy research. Currently, she serves as a research director at the Planetary Health Research Centre (PHRC), Kathmandu.

Meghnath Dhimal is an environmental health scientist with over 20 years of experience in research and policy formulation. His primary area of expertise is on environmental health, with more focus on climate change and health. Currently, he works as a senior research officer at the Nepal Health Research Council and Fellow of International Science Council.

Hom Nath Gartaula is a senior scientist at the International Rice Research Institute (IRRI) based in Los Baños Philippines. His research focuses on food security, climate change, gender and social inclusion, and agricultural innovation system.

Monika Giri, a research fellow at the Southasia Institute of Advanced Studies, is pursuing a PhD in Integrative Conservation and Environmental Anthropology at the University of Georgia, USA. Her research centers on climate and resource governance, with a particular focus on climate-induced disasters, adaptation, relocation, and just transitions.

Kirti Kusum Joshi is a scholar of urban and regional planning. He is the founding dean of the College of Engineering and IT with Lumbini Technological University, Nepal.

Kalpana Karki is promoter of women's land rights and joint land ownership and a peace activist. She worked as the campaign Coordinator at the Community Self-Reliance Centre and is also a gender focal person for the International Land Coalition and Forum Asia for strengthening the networks and alliances for women's land rights.

Rahul Karki has 15 years of work experience in research in community forestry and natural resources management. He is a trained environmentalist with practical field experience. He has published in national and international peer-reviewed journals.

Kanchan Kattel is a PhD researcher at the Department of Nutrition and Public Health, University of Agder, Norway. Her research focuses on the components of sustainable food systems and nutrition in the Himalayas.

Kanchan Lama has over 30 years of experience in the Central and South Asian region. She works as the principal investigator for the IDRC-Canada-funded project with ForestAction Nepal titled 'Economic Empowerment of Women through Forest Solutions.'

Jonathan K London is Professor of Community and Regional Development in the Department of Human Ecology at the University of California, Davis. His research and teaching focus on environment justice in rural communities and use a community-based participatory action approach.

Jony Mainaly is an independent researcher and teaches environmental law, gender, and law in Nepal. Her major area of research focuses on law, policy, justice issues governing environment, climate change, gender, and social inclusion.

Gyanu Maskey is a researcher at the South Asia Institute of Advanced Studies. Her research expertise encompasses the themes of water, forest, agriculture, and climate, with a focus on governance, justice, equity, and inclusion. She is pursuing PhD from KU School of Arts focusing on urban water security.

Dhirendra Nalbo is the co-founder of the Open Institute for Social Science, a Liberal Arts college based in Kathmandu, Nepal. He teaches and writes on peace and conflict studies primarily linked with the politics of the environmental, climate change, and geopolitics.

Madan Pariyar was born in a Dalit family at Rajbiraj in Saptari district of Nepal. He led the Government of Nepal's High Level Commission for State Restructuring. He has made notable contributions to research on inclusion and good governance and is a founding member of Samata Foundation.

Naya Sharma Paudel is a lead researcher at ForestAction Nepal. He is a political ecologist with interests in community forestry, participatory conservation, forest policy, tenure, and resource governance.

Prajal Pradhan is an assistant professor at the Energy and Sustainability Research Institute Groningen, University of Groningen, the Netherlands. His current research focuses on understanding the necessary conditions for long-term sustainability, including achieving SDGs, urban transformations, and climate resilience.

Prashanna Man Pradhan is a senior program manager at ENPHO (Environment and Public Health Organization). His expertise lies in addressing water and sanitation challenges, where he excels in water quality assessment, sanitation initiatives, and developing sustainable solutions for communities

Thomas Robertson is a historian and the former director of Fulbright Nepal/ USEF. He researches the history of development and environmental change in Nepal.

Anisha Sapkota conducts research on sustainable agricultural practices, agrobiodiversity, biodiversity conservation, and pesticides.

Poshendra Satyal serves as Senior Policy Manager at BirdLife International in Cambridge (UK). His research expertise encompasses interdisciplinary and policy-relevant studies on environment and development in Asia and Africa, with a focus on social and environmental justice within the realms of nature conservation, water governance, forest management, and climate change.

Prabindra Shakya is a human rights activist from Nepal's indigenous Newar community. He is the founder/director of the Community Empowerment & Social Justice Network (CEMSOJ) in Nepal and also the convenor of Asia Indigenous Peoples Network on Extractive Industries and Energy (AIPNEE).

Bhawana Sharma is a passionate development expert with over two decades of diverse experience in Program Management, specializing in Public Health, Sanitation, Education, and other development issues. She has worked extensively in Nepal, India, and Denmark.

Pasang Yangjee Sherpa is a Sharwa anthropologist from Pharak, the southern part of the Mount Everest region in northeastern Nepal. Her research, writing, and pedagogy focus on climate change and Indigeneity among Himalayan communities. She is an assistant professor of Lifeways in Indigenous Asia at the University of British Columbia.

Bidhya Shrestha teaches at the Central Department of Population Studies, Tribhuvan University. Her research covers gender studies, demography, and reproductive health.

Tara Lal Shrestha teaches at the Central Department of English, Tribhuvan University. His publications focus on subaltern, resistance, and justice. He also leads the Cycle Culture Community in Nepal.

Bijay Singh is a lecturer at The Institute of Engineering at Tribhuvan University, Pulchowk Campus, Nepal, with expertise in housing design, architecture, and urban planning

Sangeeta Singh is a professor in Urban Planning and Deputy Director at the Centre for Disaster Studies at the Institute of Engineering at Tribhuvan University, as well as the coordinator of the Nepal National Urban Forum representing the Institute of Engineering.

Madhusudan Subedi is a social scientist and has been working as a researcher, university teacher, and training facilitator since 1993. Currently, he is a professor and chair of the Department of Community Health Sciences, and coordinator of the School of Public Health at Patan Academy of Health Sciences, Nepal.

Mukta S. Tamang is an anthropologist and teaches at Tribhuvan University. He was a visiting fellow at Jawaharlal Nehru University, Goldsmiths College of the University of London, and worked as a research fellow at the New School for Social Research in New York.

Marissa Taylor is a writer based in Kathmandu. She has worked as a journalist in Nepal's newsrooms for a decade, with her writing focusing on the intersections of climate change, biodiversity, and public health.

Leah Temper is an ecological economist and political ecologist. She is currently the director of the Health and Economic Policy Program at the Canadian Association of Physicians for the Environment where she leads the 'Fossil Fuel Ads Make Us Sick' campaign. She is Co-Founder of the Global Atlas of Environmental Justice and the co-coordinator of the ACKnowl-EJ project (Activist-Academic co-production of Knowledge for Environmental Justice).

Sudeep Jana Thing conducts research on state-local people contestations, local resistance and social movements, human rights, equity, social justice, governance, sustainable livelihoods in relation to conservation and protected areas, as well as indigenous peoples' and local community conserved areas and territories in Nepal. He is affiliated with the Curtin University Department of Geography in the School of Design and Built Environment (Perth Australia).

Bhushan Tuladhar is Head of Environment Department and Member of City Planning Commission at Kathmandu Metropolitan City, Executive Director at Clean Energy Nepal and Environment and Public Health Organization, and Chief of Party US AID Clean Air, Nepal.

Foreword

Although it has gone by different names, the quest for environmental justice in Nepal dates back a long time. For years, and even generations, formal and informal groups have been raising their voices and organizing movements to seek justice in matters related to environmental security and livelihoods. However, since the 1950s, when the country took the path of modern development in earnest, the form and magnitude of environmental injustices and the people's concerns have changed significantly. Similarly, the reinstatement of democratic governance in 1990 led to greater openness. As a result, environmental justice concerns have changed accordingly.

In earlier times, especially prior to the 1970s, Nepal lacked modern infrastructure like roads and hydropower. The environment in those days was relatively pristine: springs, rivers, and lakes were serene and abundant with aquatic resources; forests were rich with medicinal and aromatic plants, non-timber forest products, as well as biodiversity and wildlife flourished; climatic patterns like precipitation were more or less predictable so that farmers would know about the seasons and times for planting and harvesting of crops. People's livelihood was primarily based on natural resources like agricultural land, grasslands, rivers, and forests.

In recent decades, however, the issues related to an urban way of life and urban problems have become major concerns. Climate change has become a major fear, especially for youths. It has affected primarily the marginalized classes – both economic and cultural (caste) – as they are not able to protect themselves from the adverse impacts. Accordingly, new forms of environmental injustices and environmental social justice movements have begun to appear. My own work with colleagues in environmental law over many decades has attempted to use this system to fight against environmental destruction to protect both nature and society.

Scholarship on environmental justice in Nepal had been mainly concerned with unequal access to natural resources and livelihood security because of inequity in access and distribution and due to the state's exclusive control of forests and other resources. When the ruling class noticed that declining wildlife had limited their hunting purposes and erosion and landslides threatened their lands they pushed for the conservation of these resources and enacted many draconian regulations, especially from the early 1970s. Scientific discourse on conservation that stigmatized small farmers and Indigenous people also gave impetus to enact these draconian regulations, which affected the livelihood and security of marginalized and disadvantaged people. Some progressive changes in these regulations have emerged because of strong resistance from affected communities.

As Nepal's society has changed, concerns about new environmental problems have arisen, and so have new environmental justice issues. Nepal's urbanization has expanded and deepened. Many thousands have moved from rural areas to urban areas. Some of these urban migrants include environmental refugees – those displaced from infrastructures like hydropower and also national parks. Some of them also include climate change refugees. Climate change has cut the productivity of land and other resources and fueled disasters like river cutting, floods, and landslides. These environmental problems drive people's migration to cities and market centers, where they primarily engage in the informal economy for their livelihoods, and in this process, they also get harassed by law enforcement agencies.

Given these harsh on-the-ground realities, a scholarly book on environmental justice concerns in both rural and urban areas is desperately needed. This book covers the traditional as well as emerging environmental justice issues and analyzes them in rigorous way. The succinct writing on various topics covered in various chapters is also easily readable.

The diverse chapters not only analyze problems but also offer best practices for addressing them. This is very useful in finding practical solutions for the future.

There is still no academic course or degree in environmental justice in Nepal. This book can provide important reference material if such courses are developed in Nepali colleges and universities. This book could be a much-needed resource for students, researchers, and teachers alike.

Environmental justice has become a vast academic discipline and crosscuts various traditional academic disciplines such as environmental science, environmental management/studies, sociology/anthropology, cultural studies, and legal studies. The question of 'rights', which is core in environmental justice, has been broadened to include the rights of more-than-human species including 'nature' itself.

Because of the breadth and magnitude of today's problems, it is difficult to produce an all-encompassing book on environmental justice. This is true in this book as well. Given the evolving nature of this academic field, new issues will also emerge in this field. This calls for constant research and dialog. This book will thus be useful as a foundation for future research and writing in environmental justice and for those involved in solving the problem through policies and practical actions.

Narayan Belbase

Foreword

Environmental justice is both a field of study and a field of struggle. This book, a first of its kind collection of environmental justice scholarship in Nepal, delivers on both counts.

Firstly, by compiling and synthesizing emerging scholarship on critical environmental justice studies and social movements in Nepal, and addressing issues from deforestation, air pollution, drinking water access, hydropower development, and the impacts of urbanization and climate change, it contributes to developing a uniquely Nepalese perspective on environmental justice.

Secondly, by centering the voices and experiences of Nepalese communities and amplifying the perspectives of frontline activists it serves as a catalyst for further contributing to the cohesiveness and effectiveness of environmental justice social movements in Nepal. The chapters in this book illustrate the diverse strategies and approaches employed by Nepalese activists and scholars in their quest for environmental justice, from grassroots mobilization and community-based conservation efforts to policy advocacy and legal reform, providing rich learnings for strategic considerations on advancing environmental justice policy and action in the country.

This book also enriches and expands global EJ discourse and debates. While Nepal's challenges may be context-specific, the underlying issues of power, inequality, and ecological degradation resonate with justice struggles worldwide.

In Nepal, as elsewhere, the pursuit of economic growth has resulted in ecological destruction and displacement, with vulnerable communities bearing the brunt of social and environmental costs.

As an economy still very much biomass-based, Nepal´s primary exports remain focused on wood, wool, water, and other agricultural products. In terms of tons, Nepal's biggest export may be its rivers – packaged in plastic bottles across the border to India and beyond. At the same time, the chapters show how rapid extractivist development, urbanization, and industrialization exacerbate environmental degradation and trigger place-based movements against infrastructure projects such as highways, landfills, and brick-kilns, how urban solid waste workers mobilize for health and dignity in their labor, and how peasants fight against exposure to toxic pesticides and for climate justice in Nepal.

This volume further contributes to critical issues in global environmental justice debates such as environmental casteism, provocatively questioning why caste-based inequality remains largely absent from scholarly and political discussion. It also includes chapters on worker's struggles such as the occupation-based environmental justice activism of solid waste workers and their struggle for dignity and health.

But as the authors stress, even such a broad and multi-faceted collection can only scratch the surface of people's movements for clean and healthy environments across the country. The Global Atlas of Environmental Justice, a project I helped found, similarly documents only a tiny fraction of environmental justice conflicts across the country. These include a national park where the classification of a rhino as endangered led to violent and coercive conservation against women, several conflicts over dams and a mega highway, two land acquisition conflicts for new airports, and the struggle of waste-pickers organizing for dignity, better health, and environmental labor conditions in Kathmandu.

I, therefore, join the authors in the hope that this volume inspires further writing and sharing of stories of those on the frontlines working to defend lands and protect livelihoods so as to further fill in the gaps of this rich history of environmentalism from below in Nepal. I also share the editor's hope that this book further inspires Nepalese scholars, activists, and communities to organize together to advance their solutions and alternatives.

This collection, by charting the socio-environmental history-from-below of environmental justice in Nepal, can be considered what Vijay Prashad has called socialist writing, one intended to produce a confident community of struggle for environmental justice in Nepal. I look forward to seeing how it animates further initiatives and discussions on what environmental justice in Nepal and beyond can and should look like, and how we can work to bring it about.

Leah Temper

Preface

Unlike some stories, this one has three beginnings, one middle, and an unknown ending. We offer these origin stories in our own voices.

I, Jonathan, entered the story from early educational work in Nepal, starting as a recent college graduate in the early 1990s as a volunteer with the Annapurna Conservation Area Project. At the time, ACAP was a visionary alternative to national parks that excluded people from their territories and instead attempted to integrate sustainable livelihoods and biodiversity conservation. Building on my training in community-based environmental studies, I adapted a curriculum on teaching an intergenerational oral history curriculum developed by Nancy Erbstein (later my wife) and our colleague Kristin Zimmerman to focus on cultural and ecological change in the village of Ghandruk. Here I was able to observe the complexities of community forestry and the politics of rural community development. I developed a deep appreciation of the commitment of village residents and ACAP staff to sustainable and equitable community development, while also building a critical analysis of the challenges of such initiatives.

After a long gap in educational, career, and family development, I returned to Nepal in 2015 to teach a US-Nepal learning exchange program in Nepal with Nancy and renew connections in the country. Drawing on my own decades of community-engaged research on environmental justice conflicts in California, I sought a process to develop a research agenda in Nepal that was informed by, respectful of, and beneficial to the work of Nepalese scholars, students, and activists. I have also been excited that the project has allowed me to explore the similarities and differences between EJ policy and advocacy in Nepal and in the United States. For example, Nepal's unique elements of caste discrimination, recent urbanization, and democratic governance coupled with its histories of internal colonialism stretching back millennia provide important contrasts with California and other sites in the Global North. In both countries, communities oppressed by social, political, and economic structures related to environmental and resource exploitation face dire threats to their health and well-being and in both powerful social movements have risen to confront these injustices.

Sudikshya Bhandari, a Nepalese graduate student, has served as a skillful and dedicated research and project organizer. Together we developed a comprehensive literature review of the past 25 years of environmental justice research in Nepal and conducted interviews with 10 prominent Nepalese EJ scholars to inform the outlines of our chapter and the project as a whole.

I, Jagannath, became involved in environmental justice research when, after my PhD in geography in Australia, I returned to Nepal in the 1990s eager to research social inequity and access to natural resources. A research project on this topic, supported by Nepali think-tank Martin Chautari with co-researcher Sharad Ghimire, provided the opportunity. We found that our proposed study fits well with the concept of 'environment justice (EJ)' that was being developed in the USA, South America, and India. We first developed a 'bibliography of environmental justice' and a 'glossary of keywords used in the EJ scholarship', and then we compiled relevant literature to make an EJ resource book in Nepali. This exercise helped us to understand EJ in general and the special case of Nepal, where 'environmental racism' as seen in the United States could be seen as 'environmental casteism' as lower castes are often denied access to resources underpinning their livelihoods like land and forests and were also located in hazard-prone areas.

My field research on urban environments was centered in Pokhara in collaboration with faculty in the local college. This study gave us an understanding of how the power that came with class position led to the disproportionate distribution of environmental benefits and hazards. The play of power in urban planning and management was beginning to make some locations more or less livable than others. At the time, risks due to dangerous chemicals were low due to Nepal's limited industrialization compared to cities of developed industrialized countries from where the concept of environmental racism started. But recently environmental conditions in Pokhara and other cities within Nepal have worsened, as many of the chapters in this book show.

Personally, I feel happy that I could do some work on EJ at its initial stage in Nepal. I am still using the ideas gained from this initial work in other areas like urban studies, migration and mobility, and food security. This has helped me to analyze these issues from an EJ angle. I feel very fortunate to work with many prominent scholars of the country on this project.

I, Tom, come to this volume with a background in political ecology and particularly environmental history and an interest in the on-the-ground effects of international development programs. Before this, I had done international development work as a teacher trainer in Khotang in east Nepal in the 1990s and had grown interested in what American programs had preceded me and with what effects, particularly for local groups and environments. Many years later I got a chance to research perhaps the biggest of all US programs in Nepal, its collaboration with the WHO and Nepali government on malaria eradication programs using DDT. By this time historians had started looking at international development programs but few had examined their environmental dimensions. I was particularly interested in how in Tarai places like Chitwan district social and environmental histories wove together in tangled braids; I tried to examine the malaria program and the environmental changes it brought through the eyes of variously situated Tharu, including landlords and tenants, men and women, young and old (especially old). I did archival research as well as oral histories, publishing research into this joint Nepal and American project called 'DDT and the Cold War: American Social and Environmental Engineering in the Rapti Valley (Chitwan) of Nepal.'

As an undergraduate and graduate student, I was very lucky to study with professors who in addition to teaching about history and environments also invested a great deal of time in teaching writing skills. I was originally not a fluid writer, but I learned strategies and techniques that helped make my prose more clear and compelling. Now I am someone who gets frustrated with impenetrable jargon-laden academic prose. In my view, the best academic writing tells stories of complexity, depth, and rigor without sacrificing readability. In my role as Executive Director of Fulbright Nepal in the late 2010s, I realized that even Nepal's top students and professionals faced big career obstacles because Nepali schools had not trained them to write effectively (and in some cases had taught the exact opposite of good writing). I started organizing writing workshops for Fulbright fellows and alumni. I turned those workshops into a free Nepali-language YouTube channel called 'Mitho Lekhai' and a column called 'Writing Journeys' in *The Record* featuring Nepali authors sharing experiences and tips related to non-fiction writing. We tried to work many of my professors' writing tips into the chapters of this volume.

Acknowledgments

This book's growth from an idea to the present volume has benefited greatly from many contributors for whom we are deeply grateful. We have been gratified by the generous support of our funders, including The Asia Foundation, The US Education Foundation/Fulbright Nepal, and the McConnell Foundation. Birendra Mahato, Executive Director of Community Conservation Nepal, has made his inspiring organization a crucial partner as a fiscal sponsor.

We express our appreciation for the amazing authors who have worked with us over several years to produce their excellent chapters. Finally, we dedicate this book to the next generation of scholars and activists in Nepal and beyond who are committed to informing and engaging in the struggle for environmental justice.

Jonathan K London, Jagannath Adhikari,
and Thomas Robertson
Editors

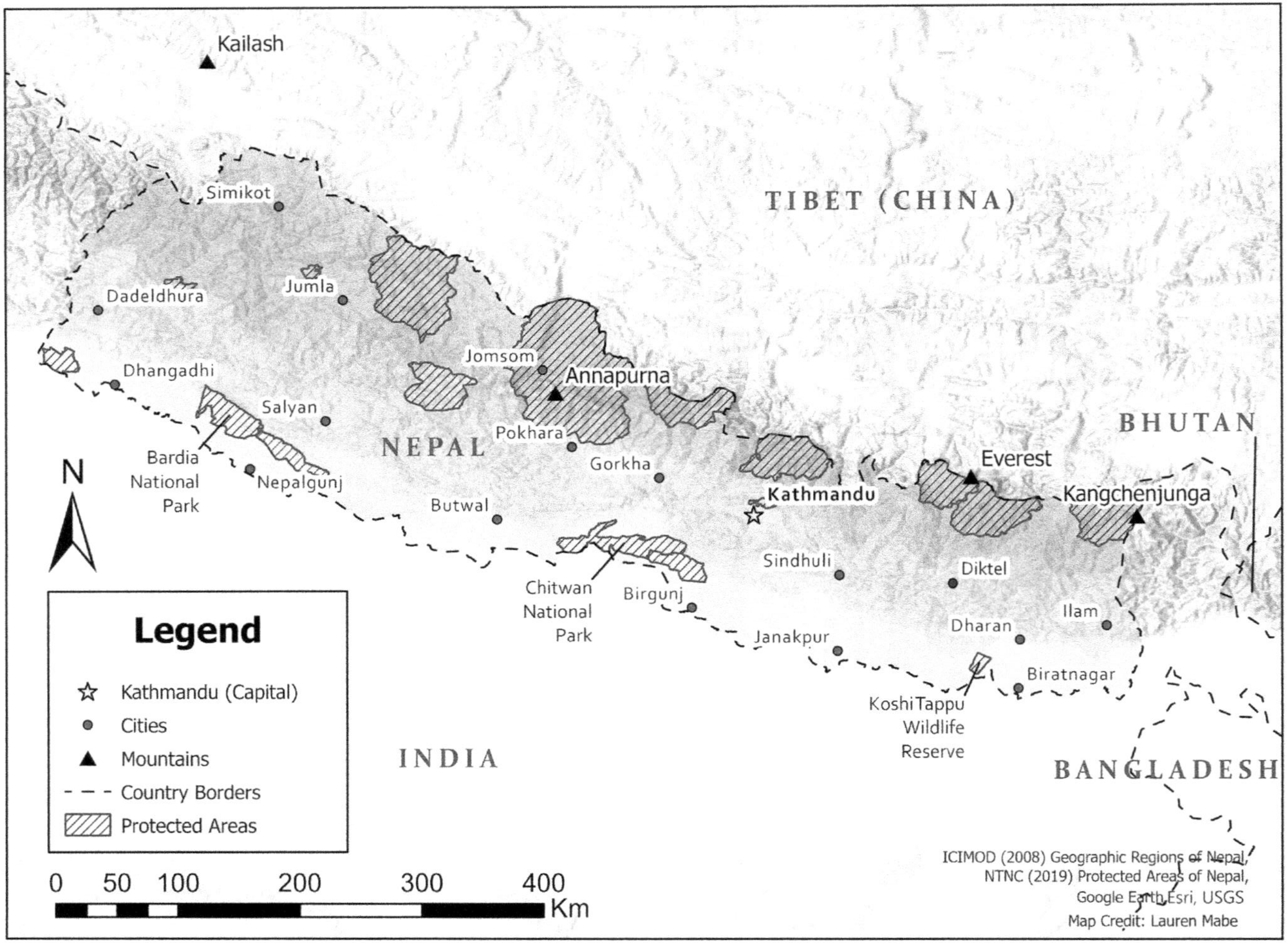

Figure 0.1 Nepal overview map.

Source: Lauren Mabe.

1 Introduction

Framing Environmental Justice Studies and Movements in Nepal

Thomas Robertson, Jonathan K London, and Jagannath Adhikari

Introduction

Nepal is a country of stunning snowcapped mountains and lush tropical forests. But like much of the Global South, the small South Asian country of 30 million residents grapples with a range of environmental threats that demand urgent attention. Deforestation, although reduced in some parts of Nepal in recent years, threatens soil erosion, habitat destruction, hydrologic degradation, and disrupted rural livelihoods. Pesticides and other chemicals undermine environmental and human health. New infrastructure such as dams, roads, and airports chew up natural areas around the country. New urban problems such as air, water, and soil pollution challenge the health of both nature and people. All of these problems are exacerbated by climate change, which is warming temperatures in the Himalayas even faster than in the rest of the planet. Nepal as a country faces climate injustice, but then its impact is more severe among its marginalized groups.

At the same time, Nepal, like much of the Global South, has a long history of profound social inequality. This exacerbates environmental problems; just as environmental problems make social marginalization even worse. Nepal is in fact one of the most unequal societies in the world. Dalits (historically known as "untouchable" occupational castes) – who make up approximately 13% of the population – face discrimination in schools, worksites, and public places as well as disparities in resource allocation. Ethnic minorities (or "tribal," "Adivasi," and "janajati" groups) – who make up approximately one-third of the population – also face discrimination and unfair resource allocation. Women of all groups face limited opportunities and inequality in all realms of private and public life. The rural and urban poor – often made up of these marginalized groups – suffer from inadequate infrastructure, lack of educational facilities, and limited access to healthcare, perpetuating the cycle of poverty. Many people face multiple forms of oppression simultaneously.

This volume looks at the overlap of Nepal's environmental problems with its deep patterns of social, economic, and political marginalization. How do environmental threats intersect with overlapping lines of caste, ethnicity, gender and social class? How do Nepal's marginalized people face disproportionately heavy burdens of pollution and environmental degradation? Given that Nepal is globally renowned

DOI: 10.4324/9781003371175-1
This chapter has been made available under a CC-BY-NC-ND 4.0 license.

for both the severity of its environmental crises and the creativity and persistence of social movements to confront them, how do marginalized groups respond to environmental problems and push for greater equality? This book stands out from others that look at individual sectors (e.g., forestry, agriculture, water resources) by combining environmental analysis with an equity-focused socio-economic analysis. That is, views Nepal's environmental and social landscapes through a holistic lens.

To do so we adopt an "environmental justice" framework. Broadly construed, environmental justice looks at how environmental inequality and social inequality are bound together in complex and mutually constitutive ways (Schlosberg 2007; Sze and London 2008; Pellow 2017). More specifically, as Jonathan London outlines in Chapter 2 of this volume, environmental injustice can be understood to be "overburdening of vulnerable populations by environmental contaminants, exclusion from environmental resources, and marginalization in environmental policy and decision-making." Nepal exemplifies cutting-edge understandings of environmental justice as simultaneously local and global and that embody struggles over customary access to land-based livelihoods as well as fights over environment quality in its growing cities.

The authors in this volume draw on a wide range of sources to define environmental justice and parse out its different strands. Several of the authors in this volume adapt Schlosberg's breakdown of multiple components of environmental justice (Schlosberg 2007, 2013) and expanded upon by others in recent decades (e.g., Almassi 2020; Whyte 2020).

1 **Distributional Justice:** This component focuses on fairness and equity and the elimination of discrimination in environmental benefits and burdens. Environmental justice calls for policies and practices that do not disproportionately impact marginalized and vulnerable communities, ensuring that everyone has an equal right to a clean and healthy environment.
2 **Participatory Justice:** Environmental justice emphasizes the meaningful involvement of affected communities in decision-making processes related to environmental policies, regulations, and projects. Inclusive participation empowers communities to advocate for their rights and have a say in shaping the environmental conditions that directly affect their lives.
3 **Recognition Justice:** This values diverse ways of knowing the environment, often used to re-center Indigenous wisdom and cosmologies as well as embodied knowledge. It also highlights the ways in which certain populations have historically shouldered inequitable burdens and carried the traces of this trauma over time.
4 **Restorative Justice:** This pro-active approach seeks to repair and compensate for past harms and injustices at multiple spatial and temporal scales, including legacies of settler colonialism, racial capitalism, and imperialism. This is often linked to a more transformative notion of reparations justice.
5 **Capabilities-based Justice:** This approach to justice focuses on building a society in which all beings can flourish and manifest their deepest possibilities.

Many chapters also illustrate Pellow's (2017) four pillars of "critical environmental justice." These employ an intersectional approach to understanding how diverse social positions produce unique experiences of environmental inequities. They take a multi-scalar view that integrates local and global levels of analysis as well as historical as well as contemporary roots of injustice. The chapters express pessimism about the capacity of the state to ameliorate environmental injustices as an ineffective champion at best and a perpetrator of such harm at worst. Finally, most view marginalized populations as indispensable to the sustainability and well-being of Nepali society and its efforts to build a sustainable and equitable future.

In this volume, we use these broad EJ frameworks to illuminate 24 richly detailed case studies from around Nepal. We take readers from climate change issues in the high mountain regions and middle hills to lead and air pollution in Kathmandu's once pristine valley to problems of deforestation and "fortress conservation" in Nepal's lowland Tarai.

It is important to note that besides London and Robertson as co-editors, all of the substantive chapters are written by Nepalese authors. This was a deliberate choice to emphasize and honor the excellent scholarship coming from Nepal and the importance of Nepalis narrating their own histories and intellectual projects. We also are pleased to feature a mix of authors in academic positions and those writing from civil society and activist organizations. This joining of writers that span the typical university-community divide is an important intervention in epistemic justice and befitting a book such as this. Likewise, the diversity of gender and cultural backgrounds of the authors has greatly enriched the book.

EJ around the World and in Nepal

Environmental justice as a concept and social movement emerged originally from the struggle for racial equality in the United States in the early 1980s. In 1982, activists in Warren County, North Carolina, protested the siting of a toxic waste facility in their predominantly black and low-income community. In the aftermath of these protests, the United States federal government requested a study of hazardous waste landfill locations and the racial and socioeconomic demographics of the surrounding communities. In 1983, the US General Accounting Office published *Siting of Hazardous Waste Landfills and Their Correlation with Racial and Economic Status of Surrounding Communities*. In 1987, the United Church of Christ (UCC) Commission for Racial Justice published a report called *Toxic Wastes and Race in the United States*, which introduced the terms environmental racism and environmental justice. Both reports documented the disproportionate siting of hazardous and toxic waste facilities near low-income communities of color. This early research argued that such siting patterns happened by a design formed through the interplay of structural racism, economic development, and environmental policy.

At the same time, the new field of political ecology had begun to emerge among scholars who studied international environmental issues. Political ecologists took issue with the strong current of Malthusian overpopulation analysis that characterized the environmental movement of the 1960s and 1970s, particularly the environmental criticism that people such as Paul Ehrlich and Garrett Hardin leveled at international development projects. Instead, political ecologists argued that many environmentalists were blaming the victim, that poor peasants of the Global South were bearing environmental hardships caused by economic exploitation and social marginalization. Key works examined famine in Nigeria, deforestation in Brazil, and indeed, soil erosion in Nepal, showing that colonial, capitalist, and otherwise exploitative economies, not ordinary villagers, had created environmental problems (Blaikie 1985; Blaikie and Brookfield 1987; Warren Dean 1995; Peet and Watts 2004).

Since these early scholarly works documented the patterns and causes of environmental injustices, the field of environmental justice studies has expanded greatly in topical focus, geographic scope, and research methodology. Sze and London (2008) refer to this as the four generations of EJ studies. Its current fourth generation is depicted usefully by Pellow's four pillars of Critical Environmental Justice noted above. Cross-cutting issues such as climate justice and Indigenous environmental justice, transboundary analyses, as well methodologies such as community-based participatory action research are rapidly transforming the field.

Environmental Justice in Nepal: *Origins, Struggles, and Prospects* is certainly not the first scholarly work on Nepal that looks at the country's environmental problems through the lens of social inequality. Jagannath Adhikari's seminal work in the early 2000s (see especially 2000, 2003) laid the foundation for this volume. Further back, Mahesh Chandra Regmi's works from the 1960s to 1980s on the economic history of Nepal, particularly his *Landownership in Nepal* (1976), include a lot of information useful for a class-based analysis of natural resource issues. Blaikie's early work (1985, 1987) on the politics of soil erosion was followed by several others who wedded social and environmental analysis. In separate studies, Stan Stevens (1993) and Ulrike Müller-Böker (1999) documented how Nepal's most famous national parks, Sagarmatha in the Everest region and Royal Chitwan in the Tarai lowlands, infringed upon the traditional rights of indigenous groups. Ramachandra Guha wrote similar critiques of Indian national parks in two influential essays (1989, 1997). Y.B. Malla (2001) examined how local elites had captured the lion's share of Nepal's famous community forestry program. R. Pandey (2004) described the toils faced by waste scavengers and street sweepers in urban Kathmandu. In 2011, Anne Rademacher (2011) showed in *Reigning the River* how river restoration of Kathmandu's Bagmati River overlapped with myopic visions of who had the right to speak for and claim river resources.

What we try to do in this book is built on this scholarship to cover more topics and add updated information. In particular, we adopt what Jonathan London (Chapter 2) refers to as a "new paradigm" for EJ studies in Nepal that integrates simultaneous historical livelihood struggles of the "environmentalism of the poor" with those related to Nepal's increasing globalization and urbanization.

Organization and Arguments of the Volume

Our first part is called Origins. In Chapter 2, "Towards a New Paradigm for Environmental Justice Studies in Nepal," Jonathan K London and Sudikshya Bhandari trace the Nepali scholarship that uses an environmental justice lens from the mid-1990s and early 2000s focus on struggles over access to land and natural resources, such as in community forestry and protected areas, to the growing diversity of the field in the late 2000s and 2010s with new topics such as water resources, sustainable agriculture, air quality, in the context of urbanization. The chapter argues that environmental justice studies in Nepal is not simply an application of theories from the Global North, or even extensions of theories from the Global South, but a distinct form that must account for Nepal's distinctive social and ecological conditions.

In Chapter 3, "People's Movements for Environmental Justice in Nepal: A Historical Perspective," Jagannath Adhikari examines the history of social movements in Nepal for their concerns about environmental issues. Examining the Rana regime in the 19th century, the Panchayat decades between 1960 and 1990, and the post-1990 years of unstable democracy, Adhikari shows that the concern of poor and marginalized in Nepal until recently has been on access to resources, particularly land, for livelihood, but that in recent years, as urbanization and industrialization has increased, social movements have begun to flag how new pollutants and hazardous substances have harmed poor, marginalized groups.

In Chapter 4, "Environmental justice and the role of Nepalese judiciary: a missed opportunity," Jony Mainali examines the recent legal history of court cases with environmental justice components. Through close analysis of several key cases, she argues that legal practitioners and scholars have mostly overlooked the potential for analysis of environmental injustice. She usefully points us toward opportunities to focus more on distributional, procedural, and restorative justice.

Part 2 begins the set of chapters on contemporary struggles in environmental justice social movements. This part focuses on Land, Forests, and Agriculture. In Chapter 5, "Environmental Injustice in Confronting Gendered Access to Land in Nepal: Joint Land Ownership as a Promising Practice," Srijana Baral, Kalpana Karki, and Kanchan Lama examine gender disparities in land access. In Nepal, women face injustice and exploitation due to lack of access to resources, particularly land. Baral, Karki, and Lama find that Nepali society's recognition of women's land rights is minimal, resulting in their exclusion from accessing financial resources, low participation in family decisions, and the lack of a welcoming atmosphere for women in land-related spaces. Conversely, the introduction of joint land ownership has proven to be a means to correct injustice in gendered land ownership patterns. Land ownership has also improved women's access to health and financial services and reduced gender-based violence.

In Chapter 6, "Environmental Justice and Unfree Agricultural Labourers in the Eastern Tarai of Nepal," Suresh Dhakal uses an environmental justice framework to analyze landlessness, social-economic inequalities, and unequal impacts of climatic hazards on the life and livelihood of a million Harawa-Charawa families, mostly Dalit bonded agricultural laborers of Nepal's Province 2 in the eastern southern plains near the Indian border. Dhakal's findings indicated that the Harawa-Charawa and other landless and marginal farmers are the ones who were harmed most by such climatic hazards.

In Chapter 7, "Connecting Dalit Land Rights with Climate Justice" Madan Pariyar and Arjun Kumar B.K. examine land rights that impede the envisioned re-distribution of lands to landless Dalit communities in Nepal and analyze how climate change impacts have harmed Dalit communities, which make up 14% of the country's population but have little political voice or access to economic resources. They explain the long and continued legacy of caste discrimination and its implications for climate justice.

In Chapter 8, "Environmental Justice and Pesticides," Kishor Athreya, Kanchan Kattel, Anisha Sapkota, and Hom Gartaula examine the pressing and growing issue of chemical pesticides in Nepal. This chapter concludes that peasants are indirectly "forced" to use chemical pesticides because of factors beyond their control, are highly exposed to toxic chemicals and unaware of the ways to protect themselves, and that the poor, marginalized and women farmers face a disproportionate burden.

In Chapter 9, "From Red to Green to Grey Hills: Reflections on the Four-Decade-Long Journey of Community Forestry and Environmental Justice in Nepal," Sunita Chaudhary writes about community forestry in Nepal, a program that has won widespread praise for helping to reforest much of the country. Chaudhary, though, points to problems such as elite capture and the limited participation of marginalized groups such as women and lower castes in what were supposed to be democratic decision-making processes. These problems have decreased participation in the last two decades, often leading to poorly managed forests and a build-up of firewood and brush, creating the conditions for a wave of forest fires.

Part 3 of the book is called "Conflicts over River and Lowland Conservation." In Chapter 10, "Protected Areas and Expendable Communities: Human-Animal Conflict Survivors and Unjust Compensation in the Koshi Tappu Wildlife Reserve," Dhirendra Nalbo shows survivors of human-wildlife conflict in Koshi Tappu Wild Reserve in Nepal's eastern lowlands face environmental injustice. Established in 1976, the KTWR experiences the single highest number of human-wildlife conflicts among Nepal's 20 Protected Areas. Nalbo argues that the environmental injustices related to compensation are linked with the historical social structure in which the government, political leaders, and elites continue to treat the economically poor and politically disenfranchised Yadav and Malaha communities as expendable "subjects" but not as the keepers and the stewards of the PAs.

In Chapter 11, "The River People and the Parks: Political Ecology of Conservation and Indigenous Livelihoods in Nepal's Terai," Naya Sharma Paudel, Sudeep Jana Thing, and Rahul Karki examine two case studies of indigenous fishing communities in and near lowland national parks – the Bote in Chitwan and the Sonaha in Bardiya. They argue that despite shifts away from protectionist paradigm toward participatory conservation toward participatory conservation, knowledge, and practices of indigenous peoples are still marginalized and undervalued. The equal footing of diverse knowledge systems and practices, productive dialogues, and partnerships between them are recommended for sustainable, just, and equitable conservation.

Part 4 of the book focuses on "Infrastructure and Indigenous Peoples." In Chapter 12, "Disaster Is Social: Uneven Effect and Recovery from the 2015

Nepal Earthquake" Mukta S. Tamang analyses the unequal consequences of the earthquake and the unequal capacity of different groups to recover. He argues that a disaster is social – a family's location in the social structure, power hierarchy, and access to various forms of economic and social capital shape its ability to anticipate and respond to catastrophe.

In Chapter 13, "Indigenous struggles for development justice in Nepal: Environmentalism on the ground," Prabindra Shakya examines how hydropower dams and roads have displaced Indigenous peoples from their houses, lands, and cultural sites. In return, they receive few development benefits while others benefit greatly. Key case studies include the Indigenous Majhi community's struggle against the Sunkoshi 2 dam project and the Kathmandu Valley Newars' resistance against the Kathmandu-Terai/Madhesh Fast Track Highway. The chapter is grounded on the understanding of the environment as a holistic system, in which Indigenous people form an important part.

Part 5 is "Urban Development and Environmental Justice." In Chapter 14, "Ensuring Health, Hygiene and Dignity for Solid Waste Workers," Prashanna Pradhan and Bhawana Sharma illustrate the plight of informal solid waste workers in Nepal as an example of occupation-based environmental injustice. It illustrates the ways in which marginalized urban migrants are stigmatized by their occupations in solid waste management which also threaten their health. The chapter profiles innovative efforts to achieve environmental justice by and on behalf of these workers through a comprehensive worker hygiene project that protects the health and dignity for these essential workers.

In Chapter 15, "Urban Environmental Justice: For Whom, From Whom?" Kirti Kusum Joshi examines the negative social and environmental impacts of narrow economic planning and analysis in two urban case studies—the Sisdol landfill site outside Kathmandu and the brick kilns of Kavre District. In both cases, Joshi discovers that standard cost–benefit economic analysis tends to overestimate economic gains and undervalue social and environmental losses, especially in regard to problems related to class, class, ethnicity, and geographic location.

In Chapter 16, "Cycling for Livelihood in Nepal: Seeking Justice on Two Wheels," Tara Lal Shrestha and Vidhya Shrestha examine livelihood-driven cyclists, their concerns in terms of risks to which they are exposed, and their agency to influence the transportation planning process in Kathmandu. Their study identifies four types of communities, whose members bicycle for their livelihoods: (i) those who use bicycles for going to the office/workplace, (ii) those who use bicycles for door-to-door hawking, (iii) those who deliver food/goods, and (iv) *kawadiwala* (those who collect reusable waste items). The authors argue that these bicycle-dependent workers are unsung environmental heroes while facing dire threats to their health and well-being.

Chapter 17, "Transport Justice on the Streets of Kathmandu" by Bhushan Tuladhar examines the problems of auto-centric urban development and transportation in Kathmandu. He depicts key decisions that led government officials to prioritize private vehicles to the detriment of pedestrians, bicyclists, and efforts to build a sustainable and equitable city. The chapter presents several case studies that highlight both innovations such as electric busses and women's bicycling classes and planning failures such as Kathmandu Valley's Ring Road.

Chapter 18 by Sangeeta Singh and Bijay Singh, "Building political capabilities through participation for environmental justice in informal housing in Kathmandu," is based on a retrospective study of the participatory approach utilized in managing informal housing. The case study is of the Kirtipur Housing Project in Kathmandu, a project to relocate residents of informal housing along the Vishnumati River corridor in 2003. The study applies an environmental justice and capabilities approach as its conceptual framework. The research emphasizes the importance of considering housing-related capabilities, encompassing social, economic, environmental, physical, and institutional aspects.

Part 6 focuses on Climate Justice. In Chapter 19, "Climate Change in Nepal through an Indigenous Environmental Justice Lens," Pasang Yangjee Sherpa examines how overly narrow biophysical, techno-managerial and apolitical approaches to climate change in Nepal's climate policies often fail to recognize how communities on the frontlines of climate change have already been adapting to climate change or to properly identify where support is needed. She shows how centering on institutional perspectives of what constitutes climate change and how to respond to it without considering local people's experiences and actions can lead to further marginalizing of Indigenous peoples and local communities.

In Chapter 20, "Women, Water and Climate: Kavre Villages in Nepal's Mid-Hills Adapt to the Increasing Impacts of the Climate Crisis," Sonia Awale, examines the impact of climate change on water availability for irrigation in the valley just to the east of the Kathmandu Valley. She finds that springs have been drying up, adding to the drudgery of women, who handle much of the agriculture because menfolk have migrated to the city or abroad for work. But, showing their resilience, women-led households continue to find ways to adapt.

In Chapter 21, "Applying a climate justice framework to understand inequities in urban water governance amid climate change challenges in Nepal," Gyanu Maskey, Poshendra Satyal, and Prajal Pradhan investigate the differential impacts of climate change on different local social groups in two Himalayan towns of Nepal – Dhulikhel and Diktel. They find that climate change and the associated impacts on the lives of people have further exacerbated the existing inequalities among the social groups – that access to drinking water for the marginalized social groups further declined because of the drying of water sources.

Part 7 looks at "Health Equity." In Chapter 22, "The Stress of Poverty on Tackling Tuberculosis in Nepal," Marissa Taylor examines the social inequities of tuberculosis, a scourge that kills 17,000 Nepalis every year. She uncovers a dangerous cycle of poverty and poor environmental conditions, forcing millions to live in overpopulated, hazardous environs with poor air quality and little hygiene. She also shows how growing drug resistance is making the problem worse.

In Chapter 23, "Impacts of Lead Contamination on Children's Health in Nepal," Meghnath Dhimal, Mandira Lamichhane Dhimal, and Madhusudan Subedi, examine the changing pattern of environmental risk factors in children in Nepal with a focus on lead contamination, which is growing because of urbanization and industrialization in Nepal. They find that increased urbanization, industrialization, municipal and electronic waste, and use of enamel paints in buildings have increased exposure to lead contamination, especially in Nepal's urban areas. Poor children and children of socially marginalized groups face a disproportionately high rate of lead exposure.

Key Lessons and Themes

1. An environmental justice approach is important for Nepal. This can illuminate the country's many intersecting social and environmental inequities, revealing the true complexity of these problems that hold the country back from sustainable and equitable development. Few are the environmental issues in Nepal that don't have a class, class, caste, or gender dimension. Few are the social issues that don't have some sort of environment dimension or consequence.
2. Nepal offers many rich case studies for global EJ scholars. Nepal offers the global EJ field rich and distinctive case studies of cutting-edge EJ issues in a developing world context. These issues are not unique but can illustrate larger patterns in valuable ways.
3. Environmental matters always have a social angle. Social inequality based on intersecting structures of class, caste, ethnicity, gender, region, disability, etc., is crucial in shaping people's differential impacts of environmental issues. Unfortunately, too much environmental analysis focuses narrowly on the technical aspects of a problem, ignoring socio-cultural factors to their peril.
4. Interdisciplinary approaches are necessary. The intersectional quality of environmental justice issues in Nepal, as elsewhere, defies traditional disciplinary boundaries. Likewise, too much technical training fails to prepare practitioners and policymakers for multi-faceted problems, and too rarely are true transdisciplinary teams created to match the complexity of real-world problems.
5. Social hierarchies have an environmental angle. Social inequality in Nepal has deep historical roots in the inequitable distribution and control of land and natural resources. These problems continue in the current day often in new forms. Social hierarchies related to class, caste, ethnicity, and gender also structure exposures to environmental hazards, creating extra burdens for those in society least able to face them. Social inequities also limit people's access to decision-making and governance.
6. Governance is important. Nepal is a young, challenged but also vibrant democracy and the EJ movements must be placed in this historical context. The issue of the roles of the state in relationship to these movements is contested and contingent. To what extent can the state address environmental and social

inequities and not exacerbate them? How can it live up to its lofty rhetoric and legal frameworks, for instance, the provision in the Constitution proclaiming that citizens are entitled "to live in a clean and healthy environment."

7 Victimization narratives must be avoided. This book attempts to move away from deterministic narratives based on class, caste, ethnicity, gender, region, and disability disparities and instead highlight efforts of resistance and transformation. We also try to not naturalize these inequities but instead show how they are historically produced and therefore can be changed.
8 Open space for nature's agency. Our stories often show how human actions led to damaging environmental outcomes, but here too, we try to avoid pure declension stories that suggest that nature has little agency or resistive capacity.
9 Resist twin problems of exotification and stigmatization. Many depictions of Nepal fall into the trap of either idealizing Nepal's spiritual and natural beauty or deriding its poverty and degradation. We follow a middle course of depicting the everyday struggles in this very real place of both peril and possibility.
10 Disaggregate data. Many of our cases show the power of governmental data disaggregated by social group and point to the need for more to illustrate social and environmental inequities and to inform transformative strategies. Disaggregated data helps us to better see social marginalization at work and provides the empirical basis for claims of injustice.
11 Listening to the voices of the socially marginalized environmental defenders. In these chapters, we try to emphasize the perspectives of those with little voice in society, although we are aware of the epistemological and methodological barriers to doing so. Where possible, we try to highlight the words and experiences of subaltern populations. Whether these are Indigenous people defending stewardship of their traditional homelands, Dalits and Janajati confronting the impacts of climate change, or women addressing legacies of patriarchal land tenure, we seek to depict this courageous agency.

At a pre-publication convening associated with the book, one speaker described environmental justice as a kind of *jerti* (glue) that can hold together strands of social-ecological analysis and social movements. We hope that the book can provide this integrative function to support Nepalese scholars and activists as they struggle to make the country live up to its democratic and social justice potential. In particular, we would be gratified to see the book used as a text in existing and new environmental justice courses and programs, a resource to inform public policy, and inspiration for environmental justice social movements in and beyond Nepal.

Despite our efforts to provide an overview of environmental justice issues and scholarship in Nepal, space and time constraints prevented the inclusion of many other important topics. We must acknowledge these as limitations of the book and call them out as fruitful new directions for future research and publications. In this forward-looking spirit, we recommend new scholarship on crucial topics such as: sustainable agriculture and food justice, in both rural and urban settings; animal rights and the rights and agency of nature; youth activism on climate change and other pressing issues; state and corporate violence against environmental defenders; environmental justice and local governance in Nepal's evolving federal system; and transboundary impacts and conflicts. Finally, we encourage new and enriched partnerships between scholars, students, activists, and policy leaders to develop and apply environmental justice research in a critical praxis of action and reflection.

References

Adhikari, J. (2000). "Urbanisation, Government Policies and Growing Social and Environmental Problems in Pokhara, Nepal." *Studies in Nepali History and Society* 5, no. 1: 43–88.

Adhikari, J. (2003). "Environmental Justice for Sustainable Livelihoods: A Grassroots Perspective." *Nepali Journal of Contemporary Studies* 3, no. 2: 29–61.

Almassi, B. (2020). *Reparative Environmental Justice in a World of Wounds*. Lexington Books.

Blaikie, P. (1985). *The Political Economy of Soil Erosion in Developing Countries*. London: Routledge.: John Wiley and Sons.

Blaikie, P. and Brookfield, H. (1987). *Land Degradation and Society*. New York/London: Routledge.

Dean, W. (1995). *With Broadax and Firebrand: The Destruction of the Brazilian Atlantic Forest*. Berkeley: University of California Press.

Guha, R. (1989). "Radical American Environmentalism and Wilderness Preservation: A Third World Critique." *Environmental Ethics* 11: 71–83.

Guha, R. (February 1997). "The Authoritarian Biologist and the Arrogance of Anti-Humanism: Wildlife Conservation in the Third World." *Ecologist* 27, no. 1: 14–20.

Malla, Y. B. (2001). "Changing Policies and the Persistence of Patron-Client Relations in Nepal: Stakeholders' Responses to Changes in Forest Policies." *Environmental History* 6, no. 2: 287–307.

Müller-Böker, U. (1999). *The Chitwan Tharus in Southern Nepal: An Ethnoecological Approach*. Stuttgart: Franz Steiner Verlag.

Pandey, R. (2004). "Solid Waste Management Practice and Health Implication: A Case of Kathmandu Metropolitan City, Nepal." *Himalayan Review 35–36*, 33–47.

Peet, R. and Watts, M. (2004). *Liberation Ecologies: Environment, Development, Social Movements*. London: Routledge.

Pellow, D. (2017). *What is Critical Environmental Justice?* New York: John Wiley & Sons.

Rademacher, A. (2011). *Reigning the River.* Durham: Duke University Press.

Regmi, M. C. (1976). *Landownership in Nepal*. Berkeley: University of California Press.

Schlosberg, D. (2007). *Defining Environmental Justice: Theories, Movements, and Nature*. Oxford: Oxford University Press.

Schlosberg, D. (2013). "Theorising Environmental Justice: The Expanding Sphere of a Discourse." *Environmental Politics* 22, no. 1: 37–55.

Stevens, S. F. (1993). *Claiming the High Ground Sherpas, Subsistence, and Environmental Change in the Highest Himalaya*. Berkeley: University of California Press.

Sze, J. and London, J. (July 2008). "Environmental Justice at the Crossroads." *Sociology Compass* 2, no. 4: 1331–1354.

United Church of Christ (UCC) Commission for Racial Justice. (1987). *Toxic Wastes and Race in the United States*. New York: UCC.

US General Accounting Office. (1983). *Siting of Hazardous Waste Landfills and Their Correlation with Racial and Economic Status of Surrounding Communities*. Washington, D.C.: US General Accounting Office.

Whyte, K. (2020). "Indigenous Environmental Justice: Anti-Colonial Action through Kinship." In B. Coolsaet (ed.), *Environmental Justice* (pp. 266–278). New York: Taylor Francis.

Part 1

Origins

2 Toward a New Paradigm for Environmental Justice Studies in Nepal

Jonathan K London and Sudikshya Bhandari

Introduction

Fierce struggles by women, low wealth, and occupational caste (Dalit) residents to gain access to local forests for fuel and fodder, urban squatters (*sukumbasis*) living near river-bank sewer outflows, children working in hazardous brick kiln factories, Indigenous peoples contesting exclusion from homelands managed as national parks, youth climate marches: These diverse struggles can all be considered instances of environmental in justice in Nepal.

Environmental justice has been defined in multiple ways since its emergence as a set of global social movements and a field of study in the late 1980s (e.g., Pulido 1996; Sze and London 2008; Pellow 2017). But, at its core, it can be understood to be the overburdening of vulnerable populations by environmental contaminants, exclusion from environmental resources, and marginalization in environmental policy and decision-making.

While malleable enough to fit a diversity of contexts, environmental justice studies must also attend to its place and time-specific characteristics. In Nepal, for example, environmental justice studies must accommodate the tremendous diversity and dynamism of the country's socio-ecological settings and structures, its layered historical legacies, and its complex positioning in regional and global political economies. In particular, EJ movements in Nepal almost always confront intersecting disparities caused by caste, class, ethnicity, and gender structures.

In the first articles on environmental justice in Nepal, Adhikari and Ghimire (2002), and Ghimire (2003) lay out an agenda for EJ studies that foreground the struggles of marginalized groups – oppressed castes, Indigenous peoples, low-wealth and others – to gain or maintain access to natural resources as a source of livelihood. Such struggles have been referred to as an "environmentalism of the poor" (Guha and Martinez-Alier 1998; Martinez-Alier 2002). These social movements confront colonial and neo-colonial states and the depredations of global capital throughout the Global South. Nepal varies somewhat from this formulation as its livelihood struggles have largely involved colonialism and imperialism from within.

Nepali contexts arguably gave rise to political ecology, an interdisciplinary field that analyzes the intersections of social processes and environmental conditions.

DOI: 10.4324/9781003371175-3
This chapter has been made available under a CC-BY-NC-ND 4.0 license.

Much of the seminal work of Piers Blaikie and colleagues in the 1980s. Blaikie's solo work (1985) and his collaborations with Cameron and Seddon (1980) and Brookfield (1987) critiqued the dominant discourse of the late 1970s and 1980s that blamed poor management by farmers in Nepal for Himalayan land degradation and soil erosion. Instead, Blaikie identified structural economic and political factors that placed "marginal people in marginal places." Blaikie identifies "underdevelopment," which he defines as poverty, inequality, and exploitation, as a result, symptom, and cause of land degradation. Because land degradation is a social, not merely environmental issue, it demands social structural solutions, namely the exploitative political economic systems.

Critiques of the "fortress conservation" model that displaced local people to create Nepal's renowned national parks also formed a conceptual framework that linked social disparities with environmental degradation (Müller-Böker 1991; Stevens 1993; McLean 1999; Mishra 1984; Robertson 2018). This critical view of how the Nepalese state has privileged bio-diversity over the health and well-being of indigenous and oppressed caste people in national parks and community forestry has continued as a core concern in Nepal's EJ studies in recent decades (Thing 2019; Ojha et al. 2022).

Adhikari and Ghimire (2002) draw on these diverse sources to differentiate EJ in Nepal and the Global South from that in the Global North, with the former focused primarily on rural livelihood struggles and the latter on urban environmental pollution. In the decade following these original articles, livelihood struggles were the central concern of Nepalese EJ studies. The increasing impacts of urbanization and globalization on Nepalese culture, economy, and politics in recent decades raises the question of whether the "environmentalism of the poor" framework is sufficient to define EJ studies in Nepal, and if not, what alternative paradigm or paradigms can incorporate the full range of phenomena.

To pursue these questions, I review the past two decades of EJ studies in Nepal to suggest that a framework that can accommodate Nepal's wide-ranging experiences with EJ problems and movements must not only include, but *integrate*, both livelihoods and environmental quality struggles as intersecting and simultaneous not sequential. In fact, it is the simultaneous and interactive character of both of these modes of struggle that make Nepal distinct from many other sites. Just as Nepali small farmers from oppressed castes and genders contend with exclusion from access to water, forests, and land in rural areas they are also subject to global economic and political forces that draw them into the cities of Nepal and counties around the world where they are subject to the onslaught of environmental pollution.

To develop this holistic framework for Nepalese EJ studies, I conducted a systematic literature review of EJ studies in Nepal (encompassing 43 peer-reviewed articles, books, and reports) as well as interviews with ten prominent Nepalese EJ scholars. The chapter is not intended to be exhaustive but does attempt to capture the temporal and topical breadth of the field. The sources were all in English and included those by Nepali scholars as well as authors from other countries.

The chapter will first sketch out some basic contours of Nepal's distinctive place and time characteristics. It will then briefly survey the past two decades of environmental justice scholarship since Adhikari's and Ghimire's early research and conclude with a synthesis of a proposed EJ paradigm fit for Nepal. In doing so, I want to acknowledge my position as a white scholar from the Global North and humbly offer my work as a contribution to the efforts of those in Nepal who are the rightful sovereigns of their own scholarship.

Placing Environmental Justice in Place and Time

Environmental justice social movements and scholarship in the United States and elsewhere in the Global North have tended to be oriented around the dual legacies of settler colonialism and slavery (Pulido 2017; Gilio-Whitaker 2019). These two core historical forces have set in motion hundreds of years of racial violence and domination, with profound social, economic, political, health, and ecological consequences.

Nepal's history has many distinct characteristics that form the foundation of environmental injustice and therefore must be the grounding for Nepalese EJ studies. These histories have been the subject of study by generations of Nepalese and foreign scholars and do not need extensive treatment here. Furthermore, many of these factors are addressed in the chapters that follow in this volume. Therefore, this chapter will merely offer a brief synthesis of some of the most prominent features.

First, unlike nearly all other countries in the Global South, Nepal was never directly colonized and therefore its indigenous peoples were not subject to the outside-driven genocide that decimated populations in settler colonial states. Instead, it was the internal colonialism of the Hindu Shah Kings that dominated Nepal's multi-ethnic populations (Whelpton 2005). Second, caste, ethnicity, and gender serve as primary axes of hierarchy and discrimination in economic, political, and cultural domains and deeply influences access to land and resources as well as environmental impacts (Levine 1987). Third, Nepal's political trends, from a monarchy (1770s–1846) to a hereditary dictatorial state and then back to the monarchy (1951), followed by a series of popular movements and the creation of a democratic, secular republic (2006) have all shaped the control of natural resources (Whelpton 2005) Nepal's current constitution (formally promulgated in 2016) contains a wide variety of social and environmental justice provisions, but legacies of caste, gender, class, and geographic discrimination make implementation an elusive aspiration (Paudel 2021). Nepal's dependence on foreign donors creates a dependency on international development institutions sapping national and popular sovereignty (Pigg 1992). Finally, Nepal's rapidly urbanizing landscapes drains and degrades the country's natural resources and imposes cumulative environmental health impacts on its population (Pradhan et al. 2020).

All of these forces drive environmental injustices with women and those from oppressed castes and classes lacking access to land and natural resources and forced into hazardous occupations and settlement conditions. And yet, this inequality does not go unchallenged as vibrant and diverse social movements continue to struggle for environmental justice.

Mapping the Contours of EJ Studies in Nepal

In their ground-breaking EJ bibliography Adhikari and Ghimire (2002), collect case studies aligned with the framework of the "environmentalism of the poor" (Guha 1989; Martinez-Alier 2002). These primarily focus on the exclusion of women, the poor, and Dalits from the benefits of community forestry and the expulsion of Indigenous peoples from areas designated as national parks. Adhikari (2003:210) terms these resource-dependent groups "eco-system people." Adhikari and Ghimire's EJ bibliography also includes urban-oriented studies such as the loss of urban green spaces, solid waste management, and *sukumbasis* in squatter settlements along urban rivers and other marginal spaces. Thus, it is not accurate to depict EJ studies in Nepal as following a linear historical trajectory from rural livelihood struggles to urban environmental injustices. Instead, a paradigm for Nepalese EJ studies has complex and multi-stranded characters from the beginning.

How should a new and integrative environmental justice paradigm be shaped? I propose that this paradigm is already in formation drawing from a synthesis of several dimensions of justice. One prominent source cited by Nepali EJ scholars is David Schlosberg's (2007, 2013) notion of the three dimensions of EJ: distributive justice, procedural justice, and recognition justice. Amartya Sen's (e.g., 2008) notion of "capabilities" or the support for the self-actualization and flourishing of individuals and communities is another important element of environmental justice in Nepal. The idea of "restorative justice" (Forsyth et al. 2021) also guides Nepali EJ studies. This is an approach that seeks to repair and compensate for past harms and injustices at multiple spatial and temporal scales, including legacies of settler colonialism, racial capitalism, and imperialism.

Many Nepali EJ scholars also credit the pioneering work on environmental racism by American EJ scholar, Robert Bullard (1990). Adhikari (2003) adapts Bullard's foundational approach to environmental racism by adding "environmental casteism" to better fit Nepal's social and religious hierarchies. More recent Nepali scholars cite David Pellow's (2017) notion of "critical environmental justice" to account for its multi-scalar, multi-temporal, and multi-species dimensions as well as its emphasis on intersectional identities and critiques of state-centric social movement strategies. The issue of inter-species justice is increasingly appearing in analyses that adopt a more-than-human or rights of nature perspective (Tschakert 2022) often associated with indigenous ways of knowing and acting (Whyte 2020).

Together, these diverse notions of justice, applied to a broad scope and scale that encompasses struggles over both livelihoods and environmental hazards, and refracted through Nepal's complex and dynamic social hierarchies are forming a vital field of environmental justice studies. While perhaps not unique, Nepal's EJ paradigm is certainly distinct enough to make crucial contributions to the field of environmental justice studies.

Resource Access and Exclusion

Scholars of EJ in Nepal have focused significant critical attention on two of the country's much-touted innovations in conservation: national parks and protected areas and community forestry. In the first case, parks have usually displayed centralized control, ostensibly for the protection of biological diversity. With community forestry, the federal government has decentralized resource access to the local scale to promote local livelihoods. In both cases, these policies have led to the exclusion of the most vulnerable populations.

Nepal's national parks have achieved global renown for protecting biodiversity. Its use of conservation areas that integrate human settlements into the landscape (such as the Annapurna Conservation Area Project) is seen as especially innovative (Gurung and Coursey 1994). However, these conservation models have also been critiqued for their monopolization of benefits by elites (Croes 2006; Schuett et al. 2016) and the exclusion of indigenous peoples from their homelands now designated as parks. Dongol and Neumann (2021) refer to this as "state making through conservation." Rai et al. (2023) apply a political ecological lens to highlight the persistence of a "fortress conversation" model, despite advances in community-based approaches. Jana Thing's extensive work in the Bardia and Chure regions of Nepal (e.g., 2019) and with colleagues (Thing et al. 2017) describes conflict between the "territories of life" that incorporate holistic cultural and economic dimensions of indigenous lifeways, from "conservation violence" that strips these communities of their land and resource tenure through physical and legal force. He calls instead for a "just conservation" approach that respects traditional forms of subsistence, values ecological knowledge, and repositions indigenous communities from the margins to the center of land and natural resource governance.

A generation of critical analyses of community forestry, building upon Guha 1991, have addressed the problems of women, the poor, and Dalits in gaining access to local forest resources (Malla 2001; Timsina 2003; Chaudhary et al. 2018; Paudel et al. 2022). Malla documents how local elites have captured most of the benefits of community forestry. Acharya and Upreti (2015) offer two conceptual frameworks through which to assess the environmental justice dimensions of environmental justice. Their "Livelihoods and Social Inclusion Framework" includes access to forest resources, agency in decision-making, and the structures of governance. They layer this onto an "Equity Framework" that can be used to assess the distribution of resources and political voice. In addition to accounts of exclusion and domination, Acharya and Upreti (2015) highlight instances of collective mobilization by these subaltern groups to claim their rights to their livelihood and social inclusion framework.

Ojha (2006) and Ojha et al. (2009) present the concept of "techno-bureaucratic doxa" – taken-for-granted governance forms – to analyze the barriers to true democratic, deliberative, and participatory forestry. Ojha et al. (2009:365) also apply the notion of "symbolic violence" to identify when claims of superior knowledge by government agencies and caste and class elites are used to exclude marginalized groups from the deliberative governance process. More recently, Ojha et al. (2022)

critique the limited success of Nepal's community forestry programs and call for a more holistic approach to ecosystem services for marginalized populations.

Paani ko Dhukkha: Suffering over Water

As in many countries, in Nepal too, access to water for irrigation and drinking in rural communities is the site of heated conflict. These conflicts have increased along with growing competition from urban and industrial uses. Many studies have identified women, Dalits, and low-income rural residents as bearing the costs of limited access to drinking water (Shrestha et al. 2020; Bhattarai et al. 2021). Forced to walk far distances to water sources makes daily livelihood a struggle: in the case of Dalits, exclusion from such sources imposes significant additional costs and risks (Vani et al. 2007). Similarly, these marginalized groups face challenges in ensuring consistent irrigation water for their crops, leading to economic and health inequities (Udas and Zwarteveeen 2005).

Clement et al. (2014) critique the apolitical and technocratic approach taken by many foreign donor-driven irrigation water programs (in particular, Integrated Water Resources Management). They note that by ignoring distributive and procedural dimensions of environmental justice such programs are unable to address systemic disparities in rural communities and often exacerbate them. A number of related articles address the inequitable impacts of large-scale water resources projects such as hydropower plants and river mining (e.g., Sikor et al. 2019, who analyze the appeal of indigenous Nepalis to global norms of justice in their opposition to such projects). These projects often disrupt rural resource-dependent livelihoods and even displace whole villages, especially those inhabited by marginalized populations (Rai 2005).

In a series of articles, Shrestha and colleagues (2018, 2019, and 2020) deploy a political ecology framework to analyze conflicts between rural communities with traditional water rights and management systems and newer peri-urban and industrial uses. These latter users degrade the traditional systems, imperiling rural livelihoods and often generating violent protests. Shrestha et al. (2018) call for greater respect for traditional water systems while helping rural areas benefit from newer economic opportunities while Shrestha (2019) critique notions of community resilience for masking local power disparities. Digging deeper into these disparities, Shrestha et al. (2020) highlights the particular challenges facing Dalits in gaining and maintaining water access.

Focusing on the city of Patan in the Kathmandu Valley, Molden et al. (2020) map the "paani ko dukkha" (suffering over water) that afflicts residents who lack access to formal, reliable, and clean water sources. This suffering is especially harsh for those of low social status such as renters, urban in-migrants without robust social networks, and those living in peripheral areas of the city. They document how the suffering over water falls most heavily on women and the poor. Related studies by Pokharel et al. (2019), and Balasubramanya et al. (2022) also emphasize the impacts of this gendered, class, and caste-based discrimination on water access.

Disparities in Urban Environmental Pollution

As Nepal urbanizes, the sites of environmental justice struggle are expanding to include its growing cities (Adhikari 2000). Numerous articles address the effects of industrial pollution on marginalized populations. Many of these studies highlight issues of gender, caste- and class-based discrimination, and disparities in environmental impacts.

One face of occupational discrimination is the fate of urban solid waste workers involved in waste pick up, transport, and disposal. Pandey (2004) focuses on the particular plight of waste scavengers and street sweepers. These workers, most often women and Dalits, face both social stigma in their association with waste as well as hazardous work conditions. Social discrimination and government neglect lead to significant health problems and early mortality. A wide range of studies have explored the failures of solid waste management in Kathmandu, highlighting faulty public policies and governance systems (Dangi et al. 2017). Few studies, however, have addressed the social disparities (such as gender, caste, or class) involved in such conflicts, suggesting an important topic for future research.

A second prime example is Sanjel et al. (2017) who profile the occupational hazards afflicting workers in Kathmandu's brick factories. Many of these workers are Dalits and rural in-migrants fleeing poverty, discrimination, and loss of livelihoods in their home villages. Sanjel et al. (2017) also document the myriad health conditions associated with inhaling brick dust as well as the air pollution spreading to nearby communities. Workers' limited political power and social marginalization reduce or even preclude advocacy for the development and enforcement of effective laws and policies for occupational health and safety. Instead, corporate firms generally get a free pass from the government to maintain unsafe working conditions and pollute local environments. Joshi and Dudani (2008) have documented the disparate impacts of brick kilns on the respiratory health of children attending school in close proximity to these facilities compared to those further away.

Several quantitative studies have sought to assess the socio-economic factors undermining air quality and other environmental regulations. Gurung et al. (2017) showed strong correlations between air quality and hospital admissions data in Kathmandu Valley but did not find significant differences by age, sex, or socio-economic status. They do, however, acknowledge that limitations in data may obscure possible effects and call for additional research. Similarly, using self-reported environmental quality perceptions, Flacke et al. (2022) found spatial, but not social, disparities in perceptions of air and water quality across six Kathmandu neighborhoods. As in the case of the solid waste management literature presented above, these studies demonstrate the need for additional research on urban environmental justice issues, including the crucial factors of caste and ethnicity.

Urban development has provoked a number of land-use conflicts. One of the most vexed is that of homelessness and housing insecurity, represented in the case of *sukumbasi* or squatter communities in many urban centers. Now living in urban settings, these landless residents are typically rural villagers displaced by the loss of agricultural livelihoods – often flooding and other natural disasters increasingly due to climate-related stressors. These communities both suffer from significant environmental harms such as exposure to air and water pollution and food insecurity and have also been blamed for contributing to water pollution due to limited sanitation facilities. Their settlements are often razed as part of urban development programs and they have little political voice to protect their health and well-being (Rademacher 2011; Ninglekhu 2017).

Another symptom of unregulated urban development is the loss of public open or green space. This deprives city dwellers of spaces for recreation and relief from urban noise and sight pollution (Adhikari 2000). This loss is especially felt in the dense urban neighborhoods inhabited by lower-income people who do not have private spaces for sports and recreation (Adhikari 2004). The increasing development of gated communities is exacerbating these disparities.

Going Global

Climate change is an existential threat to all Nepalis. However, it hits those with the least capacity to buffer themselves from its impacts and who are often marginalized from adaptation policies and programs.

A growing literature on climate justice in Nepal addresses gender and ethnicity-related disparities (e.g., Onta and Resurreccion 2011). Pandey et al. (2021) document how women are disproportionally harmed by climate change and are excluded from adaptation programs. Bhattarai (2020) links gender discrimination to problems of climate change resilience in community forestry while Bhattarai et al. (2015) apply political ecology and socio-ecological systems thinking to focus on the gendered disparities in agro-biodiversity management. Addressing ethnic, epistemic, and spatial disparities in Nepal's Everest region, Pasang Sherpa has produced a series of important articles (e.g., 2014, 2015, and 2021) that highlight alternative visions of centering local knowledge among Sherpa and other *janajati* ethnicities in the high Himalaya in confronting climate disruption. Climate justice is not only a rural phenomenon in Nepal, as shown by Giri et al. (2021), who assess how residents in urban informal settlements face severe climate impacts. Climate justice is sure to be a growing element of environmental justice studies as conditions and disparities worsen.

Conclusion

As demonstrated above, Nepal represents a vibrant and growing edge in the global field of environmental justice studies. Its literature spans spatial, issue, and demographic boundaries and applies a diversity of theoretical lenses to its complex socio-ecological challenges. From the beginning, it has both drawn from classic EJ literature (e.g. Bullard 1990; Schlosberg 2007), political ecology (Blaikie 1985; Blaike and Brookfield 1987), feminist studies (Rocheleau et al. 1996), and critical analysis of caste and ethnicity (Adhikari 2000; Bishwakarma 2019). It analyzes the depredations of the development state and neo-liberal capital, although the latter theme is arguably less well-developed than the first. Applying a gender, caste, and class analysis to urban areas is another important growing edge.

The recent literature on environmental studies in Nepal has not so much departed from its roots in the environmentalism of the poor but integrated this into a critical framework that can accommodate both urban-based-environmental disparities as well as the complex rural/urban nexus. It is the simultaneous and mutually constitutive character of the rural and urban, the local and the global, and the struggle over livelihoods and against environmental hazards that makes Nepal distinct in the broader field of EJ studies. While still forming its own distinct and cohesive identity as a field, EJ studies in Nepal represent a crucial site of scholarly innovation and deserve careful attention from global EJ academic and activist networks.

References

Acharya, S., & Upreti, B. R. (2015). "Equity, Inclusion and Conflict in Community Based Forest Management: A Case of Salghari Community Forest in Nepal." *Dhaulagiri Journal of Sociology and Anthropology*, *9*, 209–223.

Adhikari, J. (2000). "Urbanisation, Government Policies and Growing Social and Environmental Problems in Pokhara, Nepal." *Studies in Nepali History and Society*, *5*(1), 43–88.

Adhikari, J. (2003). "Environmental Justice for Sustainable Livelihoods: A Grassroots Perspective." *Nepali Journal of Contemporary Studies*, *3*(2), 29–61.

Adhikari, J. (2004). "A Socio-ecological Analysis of The Loss of Public Properties in an Urban Neighborhood: A Case Study of Pokhara, Nepal." *Contributions to Nepalese Studies*, *31*(1), 85–114.

Adhikari, J., & Ghimire, S. (2002). *Bibliography on Environmental Justice in Nepal*. Martin Chautari Centre for Social Research and Development.

Balasubramanya, S., Stifel, D., Alvi, M., & Ringler, C. (2022). "The Role of Social Identity in Improving Access to Water, Sanitation and Hygiene (WASH) and Health Services: Evidence from Nepal." *Development Policy Review*, *40*(4), e12588.

Bhattarai, B. (2020). "How Do Gender Relations Shape a Community's Ability to Adapt to Climate Change? Insights from Nepal's Community Forestry." *Climate and Development*, *12*(10), 876–887.

Bhattarai, B., Beilin, R., & Ford, R. (2015). "Gender, Agrobiodiversity, and Climate Change: A Study of Adaptation Practices in the Nepal Himalayas." *World Development*, *70*, 122–132.

Bhattarai, B., Upadhyaya, R., Neupane, K. R., Devkota, K., Maskey, G., Shrestha, S., … & Ojha, H. (2021). "Gender Inequality in Urban Water Governance: Continuity and Change in Two Towns of Nepal." *World Water Policy*, *7*(1), 30–51.

Bishwakarma, M. (2019). *Political Transformations in Nepal: Dalit Inequality and Social Justice*. Routledge.

Blaikie, P. (1985). *The Political Economy of Soil Erosion in Developing Countries*. London: Routledge.

Blaikie, P., & Brookfield, H. (1987). *Land Degradation and Society*. Routledge.

Blaikie, P. M., Cameron, J., & Seddon, D. (1980). *Nepal in Crisis: Stagnation and Growth at the Periphery*. Oxford: Oxford University Press.

Bullard, R. D. (1990). *Dumping in Dixie: Race, Class, and Environmental Quality*. Routledge.

Chaudhary, S., McGregor, A., Houston, D., & Chettri, N. (2018). "Environmental Justice and Ecosystem Services: A Disaggregated Analysis of Community Access to Forest Benefits in Nepal." *Ecosystem Services*, *29*, 99–115.

Clement, F., Basnet, G., Sugden, F., & Bharat, L. (2014). "Social and Environmental Justice in Foreign Aid: A Case Study of Irrigation Interventions in Western Nepal." *New Angle: Nepal Journal of Social Science and Public Policy*, *3*(1), 65–83.

Croes, K. D. (2006). "Conserving the King: Inverting the Origin Story of the Annapurna Conservation Area Project of Nepal." *Himalaya, The Journal of the Association for Nepal and Himalayan Studies*, *26*(1), 6.

Dangi, M. B., Schoenberger, E., & Boland, J. J. (2017). "Assessment of Environmental Policy Implementation in Solid Waste Management in Kathmandu, Nepal." *Waste Management & Research*, *35*(6), 618–626.

Dongol, Y., & Neumann, R. P. (2021). "State Making through Conservation: The Case of Post-Conflict Nepal." *Political Geography*, *85*, 102327.

Flacke, J., Maharjan, B., Shrestha, R., & Martinez, J. (2022). "Environmental Inequalities in Kathmandu, Nepal: Household Perceptions of Changes between 2013 and 2021." *Frontiers in Sustainable Cities*, *4*, 1–19.

Forsyth, M., Cleland, D., Tepper, F., Hollingworth, D., Soares, M., Nairn, A., & Wilkinson, C. (2021). "A Future Agenda for Environmental Restorative Justice." *The International Journal of Restorative Justice*, *4*(1), 17–40.

Ghimire, S. (2003). "Concept of Environmental Justice in Nepal: Environmentalism of the Poor for Sustainable Livelihood." *Himalayan Journal of Sciences*, *1*(1), 47–50.

Gilio-Whitaker, D. (2019). *As Long as Grass Grows: The Indigenous Fight for Environmental Justice, from Colonization to Standing Rock*. Beacon Press.

Giri, M., Bista, G., Singh, P. K., & Pandey, R. (2021). "Climate Change Vulnerability Assessment of Urban Informal Settlers in Nepal, a Least Developed Country." *Journal of Cleaner Production*, *307*, 127213.

Guha, R. (1991). *The Unquiet Woods: Ecological Change and Peasant Resistance in the Himalaya*. Oxford: Oxford University Press.

Guha, R. and Martinez-Alier, J. (1998). *Varieties of Environmentalism: Essays North and South*. London: Earthscan.

Gurung, A., Son, J. Y., & Bell, M. L. (2017). "Particulate Matter and Risk of Hospital Admission in the Kathmandu Valley, Nepal: A Case-Crossover Study." *American Journal of Epidemiology*, *186*(5), 573–580.

Gurung, C. P., & Coursey, M. D. (1994). "The Annapurna Conservation Area Project: A Pioneering Example of Sustainable Tourism?" In E. Cater and G. Lowman, *Ecotourism: A Sustainable Option?* Chichester: Wiley. 177–194.

Joshi, S. K., & Dudani, I. (2008). "Environmental Health Effects of Brick Kilns in Kathmandu Valley." *Kathmandu University Medical Journal (KUMJ)*, *6*(1), 3–11.

Levine, N. E. (1987). "Caste, State, and Ethnic Boundaries in Nepal." *The Journal of Asian Studies*, *46*(1), 71–88.

Malla, Y. B. (2001). "Changing Policies and the Persistence of Patron-Client Relations in Nepal: Stakeholders' Responses to Changes in Forest Policies." *Environmental History*, *6*(2), 287–307.

Martinez-Alier, J. (2002). *The Environmentalism of the Poor: A Study of Ecological Conflicts and Valuation*. Edward Elgar Publishing.

McLean, J. (1999). "Conservation and the Impact of Relocation on the Tharus of Chitwan, Nepal." *Himalaya, the Journal of the Association for Nepal and Himalayan Studies*, *19*(2), 8.

Mishra, H. R. (1984). "A Delicate Balance: Tigers, Rhinoceros, Tourists and Park Management vs. The Needs of the Local People in Royal Chitwan National Park, Nepal." In J. A. McNeely & K. Miller (Eds.), *National Parks, Conservation, and Development: The Role of Protected Areas in Sustaining Society*, 197–205. Smithsonian Institution Press.

Molden, O. C., Khanal, A., & Pradhan, N. (2020). "The Pain of Water: A Household Perspective of Water Insecurity and Inequity in the Kathmandu Valley." *Water Policy*, *22*(S1), 130–145.

Müller-Böker, U. (1991). "Wild Animals and Poor People: Conflicts between Conservation and Human Needs in Citawan (Nepal)." *European Bulletin of Himalayan Research*, *2*, 28–31.

Ninglekhu, S. (2017). "'Inauthentic' Sukumbasi: The Politics of Aesthetics and Urgency in Kathmandu." *Himalaya*, *37*(1), 72–83.

Ojha, H., Paudel, N. S., Timsina, J., Chaudhary, S., & Baral, H. (2022). "Ecosystems Services from Community Forestry: Prospects and Challenges for Improving Local Livelihoods in Nepal." In J. Timilsina, T. Marseni, D. Gauchan, J. Adhikari, H. Ojha (Eds.) *Agriculture, Natural Resources and Food Security: Lessons from Nepal*, Cham, Switzerland: Springer, 337–356.

Ojha, H. R. (2006). "Techno-bureaucratic Doxa and Challenges for Deliberative Governance: The Case of Community Forestry Policy and Practice in Nepal." *Policy and Society*, *25*(2), 131–175.

Ojha, H. R., Cameron, J., & Kumar, C. (2009). "Deliberation or Symbolic Violence? The Governance of Community Forestry in Nepal." *Forest Policy and Economics*, *11*(5–6), 365–374.

Onta, N., & Resurreccion, B. P. (2011). "The Role of Gender and Caste in Climate Adaptation Strategies in Nepal." *Mountain Research and Development*, *31*(4), 351–356.

Pandey, A., Prakash, A., & Werners, S. E. (2021). "Matches, Mismatches and Priorities of Pathways from a Climate-Resilient Development Perspective in the Mountains of Nepal." *Environmental Science & Policy*, *125*, 135–145.

Pandey, R. (2004). "Solid Waste Management Practice and Health Implication: A Case of Kathmandu Metropolitan City, Nepal." *Himalayan Review 35–36*, 33–47.

Paudel D. (2021) Social Inclusion, Sustainable Development, and ASTA-JA in Nepal, In M. Bhandari & S. Hanna (Eds.) *Inequality-The Unbeatable Challenge*, New York: River Publishers.

Paudel, G., Carr, J., & Munro, P. G. (2022). "Community Forestry in Nepal: A Critical Review." *International Forestry Review*, *24*(1), 43–58.

Pellow, D. (2017). *What is Critical Environmental Justice?* New York: John Wiley & Sons.

Pigg, S. (1992). "Inventing Social Categories through Place: Social Representations and Development in Nepal." *Comparative Studies in Society and History*, *34*(3), 491–513.

Pokharel, K., Joshi, T., Adhikari, J., & Neupane, K. R. (2019). "Water Security in Urban Water Governance: A Case Study of Dhulikhel Municipality, Kavrepalanchowk, Nepal." *New Angle: Nepal Journal of Social Science and Public Policy*, *5*(1), 32–48.

Pradhan, B., Sharma, P., & Pradhan, P. (2020). Urban Growth and Environment and Health Hazards in Kathmandu Valley, Nepal in R. Singh, B. Srinagesh, & S. Anand (Eds.) *Urban Health Risk and Resilience in Asian Cities*. 293–324. Singapore: Springer.

Pulido, L. (1996). "A Critical Review of the Methodology of Environmental Racism Research." *Antipode*, *28*(2), 142–159.

Pulido, L. (2017). Geographies of race and ethnicity II: Environmental racism, racial capitalism and state-sanctioned violence. *Progress in Human Geography 41*(4), 524–533.

Rademacher, A. (2011). *Reigning the River*. Duke University Press.

Rai, I. M., Melles, G., & Gautam, S. (2023). "Community Development for Bote in Chitwan National Park, Nepal: A Political Ecology of Development Logic of Erasure." *Sustainability, 15*, 2834.
Rai, K. (2005). *Dam Development: Dynamics of Social Inequality in a Hydropower Project in Nepal*. Cuvillier Verlag.
Robertson, T. B. (2018). "DDT and the Cold War Jungle: American Environmental and Social Engineering in the Rapti Valley of Nepal." *Journal of American History, 104*(4), 904–930.
Rocheleau, D., Wangari, E., & Thomas-Slayter, B. (1996). *Feminist Political Ecology: Global Issues and Local Experience*. Routledge.
Sanjel, S., Khanal, S. N., Thygerson, S. M., Carter, W. S., Johnston, J. D., & Joshi, S. K. (2017). "Respiratory Symptoms and Illnesses Related to the Concentration of Airborne Particulate Matter among Brick Kiln Workers in Kathmandu Valley, Nepal." *Annals of Occupational and Environmental Medicine, 29*, 1–12.
Schlosberg, D. (2007). *Defining Environmental Justice: Theories, Movements, and Nature*. Oxford University Press.
Schlosberg, D. (2013). "Theorising Environmental Justice: The Expanding Sphere of a Discourse." *Environmental Politics, 22*(1), 37–55.
Schuett, M. A., Dahal, S., & Nepal, S. (2016). "Local Perspectives on Benefits of an Integrated Conservation and Development Project: The Annapurna Conservation Area in Nepal." *International Journal of Biodiversity and Conservation, 8*(7), 138–146.
Sen, A. (2008). "The Idea of Justice." *Journal of Human Development, 9*(3), 331–342.
Sherpa, P. (2014). "Climate Change, Perceptions, and Social Heterogeneity in Pharak, Mount Everest Region of Nepal." *Human Organization, 73*(2), 153–161.
Sherpa, P. Y. (2015). "Institutional Climate Change Adaptation Efforts among the Sherpas of the Mount Everest Region, Nepal." In D. Wood (Ed.) *Climate Change, Culture, and Economics: Anthropological Investigations*. U.K.: Emerald Group Publishing Limited. 1–23.
Sherpa, P. Y. (2021). "Nepal's Climate-Change Cultural World." In P. Sillitoe (Ed.), *The Anthroposcene of Weather and Climate: Ethnographic Contributions to the Climate Change Debate*. 220–245.
Shrestha, A. (2019). "Which Community, Whose Resilience? Critical Reflections on Community Resilience in Peri-Urban Kathmandu Valley." *Critical Asian Studies, 51*(4), 493–514.
Shrestha, A., Joshi, D., & Roth, D. (2020). "The Hydro-Social Dynamics of Exclusion and Water Insecurity of Dalits in Peri-Urban Kathmandu Valley, Nepal: Fluid Yet Unchanging." *Contemporary South Asia, 28*(3), 320–335.
Shrestha, A., Roth, D., & Joshi, D. (2018). "Flows of Change: Dynamic Water Rights and Water Access in Peri-Urban Kathmandu." *Ecology and Society, 23*(2), 42.
Sikor, T., Satyal, P., Dhungana, H., & Maskey, G. (2019). "Brokering Justice: Global Indigenous Rights and Struggles over Hydropower in Nepal." *Canadian Journal of Development Studies/Revue canadienne d'études du développement, 40*(3), 311–329.
Stevens, S. F. (1993). *Claiming the High Ground Sherpas, Subsistence, and Environmental Change in the Highest Himalaya*. University of California Press.
Sze, J. and London, J. (July 2008). "Environmental Justice at the Crossroads." *Sociology Compass, 2*(4), 1331–1354.
Thing, S. J. (2019). "Politics of Conservation, Moral Ecology and Resistance by the Sonaha Indigenous Minorities of Nepal." In C. Griffin, R. Jones, I. Robertson (Eds.), *Moral Ecologies: Histories of Conservation, Dispossession and Resistance*, Cham, Switzerland: Palgrave Macmillan, 37–58.

Thing, S. J., Jones, R., & Jones, C. B. (2017). "The Politics of Conservation: Sonaha, Riverscape in the Bardia National Park and Buffer Zone, Nepal." *Conservation and Society, 15*(3), 292–303.

Timsina, N. P. (2003). "Promoting Social Justice and Conserving Montane Forest Environments: A Case Study of Nepal's Community Forestry Programme." *Geographical Journal, 169*(3), 236–242.

Tschakert, P. (2022). "More-than-Human Solidarity and Multispecies Justice in the Climate Crisis." *Environmental Politics, 31*(2), 277–296.

Udas, P. B., & Zwarteveen, M. (2005). "Prescribing Gender Equity? The Case of the Tukucha Nala Irrigation System, Central Nepal." In D. Roth, R. Boelens, M. Zwarteveen (Eds.), *Liquid Relations: Contested Water Rights and Legal Complexity*, New Brunswick: Rutgers University Press, 21–43.

Vani, M. S., Asthana, R., Belbase, N., Thapa, L. B., Pastakia, F., & Moore, P. (2007). *Environmental Justice and Rural Communities: Studies from India and Nepal*. IUCN.

Whelpton, J. (2005). *A History of Nepal*. Cambridge University Press.

Whyte, K. (2020). "Indigenous Environmental Justice: Anti-colonial Action through Kinship." In B. Coolsaet (Ed.), *Environmental Justice: Key Issues*, 266–278. Routledge.

3 Environmental Justice Movements in Nepal

A Historical Perspective

Jagannath Adhikari

Introduction

This chapter analyses the environmental justice concerns of people in Nepal's history in three periods: the days of national unification and dictatorial rule of the Ranas (1743–1951), the downfall of the Rana regime to the end of the Panchayat period from 1950 to 1990, and the current democratic system and its variations from 1990 onward. These were major changes, but they do not mark a point of total rupture in the political and socio-economic system. Many social structures, such as the caste system and patriarchy, persisted across different political eras. Nepal's environmental justice movements vary in different periods of the country's history because of changing social-political structures, knowledge systems, and technological developments.

This chapter aims to identify and analyze these concerns and collective actions – both passive and active – based mainly on a literature survey and supplemented by oral histories drawn from marginalized communities. Two case studies from the 1990s – one about the bonded-labor (Kamaiya Tharu) liberation movement and another about an urban ecological movement in Pokhara – are also discussed to illustrate the change in people's concerns and in the process of organizing environmental justice movements.

Looking at environmental justice movements worldwide, Sicotte and Brulle (2017) and Auyero and Swistun (2009) argue that collective action, as a response to injustice is relatively uncommon for people who have been subjected to long-term material and political hardships. As a result, it is often difficult to track such actions, especially those that occurred in the past when overt expression was difficult because of repressive political regimes. In such cases, the existence of any form of social resistance could indicate that something beyond miserable conditions like poor housing conditions must have provoked people to mobilize. As this chapter shows, this is the case in Nepal especially prior to the rise of democracy in 1990, as the repressive political systems then did not allow collective action or resistance against the government. Hence, whatever concerns were expressed in those repressive regimes indicate larger problems for the people.

DOI: 10.4324/9781003371175-4
This chapter has been made available under a CC-BY-NC-ND 4.0 license.

National Unification to the End of Rana Regime (1743–1951)

Prior to the consolidation of Nepal's current borders in the 1760s and 1770s, the central Himalaya was home to about 80 micro-states known as *baise* and *chaubise* kingdoms. Not much is known about the environmental issues in these micro-states; some studies point to the fact that local chieftains were responsive to their subject's material concerns. For example, Gorkha, one of those micro-states, was known for supporting a socially-just model of environmental and natural resource conservation (Paudyal et al. 2022). At that time, environmental resources were managed through long-established place- and ethnically-specific systems (Gurung 1997; Fisher 1994; Nesheim 1992).

During the Rana period, common people faced severe hardships because of the extraction of surplus wealth through taxes, and exploitation through unpaid work, compulsory labor for the creation of infrastructure, and the entertainment of ruling families, including through local revenue functionaries. These functionaries were usually the village landlords and had jurisdiction over different administrative units as well as different forms of unpaid labor variously known as *begari*, *jhara* and *hulak* (Regmi 1971). The high-ranking families and those serving the government were provided with land called *Birta* and *Jagir*. Peasants had to pay half of their produce, which meant all surpluses would be taken away. Here is an account by Francis Hamilton, a British visitor in the districts of Bara, Parsa, and Rautahat:

> The peasantry are extremely nasty, and apparently indigent. Their huts are small, dirty and very ill calculated to keep out the cold winds of the winter season, for a great many of them have no other walls, but a few reeds supported by sticks in a perpendicular direction. Their clothing consists of some cotton rags, neither bleached nor dyed, and which seem never to be washed.
> (Hamilton 1819; as quoted in Regmi 1999, 169)

The people described by Hamilton were probably tribal Tharus. Their condition at that time was affected by the government policies and the practices of the ruling class, especially the hunting of large game animals for sport. Robertson (2021) describes the mammoth scale and grandeur of such hunting trips of Nepali and British royal families and Rana rulers, which put a huge burden on Tharus. Such trips started mainly from the late 1850s and reached a peak in the 1910s and 1920s. They were organized in the name of British-Nepal diplomacy or in the name of helping people to kill predatory wild animals. In all cases, it was the Tharus who suffered while the elites prospered and enjoyed their hunts in the traditional Tharu homeland.

It was tragic that local people like Tharus – who were generally good stewards of their lands and waters – were forced to destroy their ecosystem for elites but were not allowed to use it for their welfare. For example, they did not build good houses even by using the resources they had like trees because of fear that government administrators or agents of government would ask for heavy taxes or confiscate their houses. Similarly, they had to shoulder the responsibility for arranging the logistics for hunting and touring trips of the ruling class, for which they had to donate labor as well as food like goats and good quality rice.

Even in the face of such exploitation, there was little prospect for political mobilization or an active movement because of the repressive regime. On the other hand, there were instances of passive resistance or what Scott (1985) describes as 'weapons of the weak'. Tree/timber theft, poaching, and forest fire in the forests of Terai were evidence of these 'weapons of the weak' resistance. These activities in Terai frustrated the Rana state's desire to control or have a monopoly on hunting/shooting, and timber export. Avash Bhandari[1] refers that in a letter [Nepal Residency File No. 4497 of 1909] Chandra Shamsher wrote to the then British Resident Macdonald in 1909 about Ban Gujjars and their "pernicious habit of destroying forests to feed their cattle." Chandra wrote that he ordered his officials to "stop it by driving them out of our forests." There were also cases of thefts/smuggling of logs stacked for sale in the Nepali forests that seem to have perturbed the Rana officials. The Rana government sold a large quantity of timber to the British-India government, but restricted ecosystem people like Ban Gujjars from entering into forest and using forest products for their basic livelihood.

The main concern of the marginalized people in Nepal during Rana rule was to have access to land and other natural resources such as forest and water. Most people depended on primary resources for their livelihood. However, the ruling class allocated a large proportion of farmland to a few high-ranking officials, their own family members, and army officials as land grants like *Birta* and *Jagir*. The lack of access to land – a problem compounded by high taxes – meant a lot of these ordinary people found themselves in economic trouble. Some of them were evicted from their land and a large proportion of them immigrated to India – mainly to the North-East Hills, where they worked as cattle herders, dairy farmers, and laborers to clear the forest, to work in tea farms or in coal mines (Regmi 1999).

Like the oppressed Tharu's passive resistance, some other groups were also found to resist the state's extractive activities. For example, when the Nepali government interfered with the traditional land ownership and management system of Kipat in eastern Nepal, people actively revolted in 1893, halting the government's revenue survey project (Caplan 2000, 175).

The government as early as the 1850s was concerned with environmental problems, especially the cutting of trees on farmland, forestland, and public lands. Regmi (2002, 221–222) states that people were prohibited from cutting trees on public land and Jagir land (land granted to government employees) and in areas close to rivers, water sources, and irrigation channels. However, Birta landowners (generally the members of the ruling class) were allowed to cut trees.

The Ranas promulgated similar forest laws in the late 1880s. Given that the government clear cut the hardwood forests of *sal* (*Shorea robusta*) on a massive scale and sold those to British-India companies leading to deforestation (Adhikari and Dhungana 2010), the restrictions on local people cutting trees indicates a clear class discrimination.

Urban Environmental Struggles

Historical accounts (Regmi 2002, 240–224) reveal that Kathmandu under the Ranas was polluted due to garbage dumping and other unsanitary practices. Regulation in this regard was developed in 1902 regarding the household's role to collect garbage (Regmi 2002, 240–241). These regulations were created because pollution also affected the areas where the ruling class lived. There is no information about how and where the garbage was disposed of and how it affected marginalized groups. Nevertheless, some marginalized caste people like *pode* and *chayme* were assigned these sanitary tasks as higher caste people would not engage themselves in such works – perceived by them as physically and culturally polluting.

The Panchayat Period (1960–1990)

After the downfall of the Rana regime, Nepal's government started the process of planned development and modernization. Private forests were nationalized in 1957 through the Private Forests Nationalization Act. This Act, though aimed to seize the forestland gifted to Rana and elite families in the form of *Birta*, led to the perception over the decades that the state was the exclusive agency to protect forests from the people through the use of force and coercion. This is considered the cause of the widespread forest destruction that followed. This policy led people to think that the forests they protected would be taken by the government barring them from using them (e.g., Gilmour and Fisher 1991). Hence, they did not protect the forest.

During the Panchayat period, many protected areas were declared, commencing with a rhino sanctuary in Chitwan, which eventually transformed into Royal Chitwan National Park in 1973. The concept of the national park was developed from a collaborative work of foreign experts and Nepali ruling class and elites for the protection of both flora and fauna, which called for evictions of people living within the park boundaries. By 1990, seven national parks, one wildlife reserve, and one hunting reserve had been created which caused displacement of people who had been living in the areas for generations (Karki 2022).

This displacement caused a significant human toll. For example, when Rara National Park was established in 1976, 250 families from this high mountain temperate environment were resettled in the tropical Terai region (Shah 1991). Forced from their homes, even pregnant women, the elderly, and children had to walk for weeks to reach their new homes, and many died, as they could not adapt to the hot weather. Some people also returned to the Rara area, becoming landless and homeless in their former homeland.[2]

It was a time of monarchical dictatorship; overt protest was not possible. The author had a chance to talk to the warden involved in the creation of Rara Park. He said, 'I had to strictly follow the instructions from the higher authorities in this regard and had to turn blind eyes to the peoples' cries'. In the displaced villages, satirical words were developed to show the plight of the people. For example, some residents called this beautiful Tal (lake) Kal (death) and 'rangers' 'danger'.

During this period, Nepal's environmental policies were also influenced by alarmist external studies. The pessimistic forecast of such studies (e.g., Eckholm 1976) and the Grand Theory of Himalayan degradation (Ives and Messerli 2003) based on such studies blamed deforestation and other environmental problems on population pressure without recognizing that local people were also able to conserve the environment given the chance. These theories would later be contested by political ecologists such as Piers Blaikie (e.g., 1983).

After two decades of failure, the government realized that central bureaucratic forest administration control was in fact causing deforestation. In response, it promulgated the *Panchayat Forest* and *Panchayat Protected Forest Rules* in 1978, giving responsibility to local communities to manage the local forests. Because these local political units were controlled by local elites supporting the regime, the actual forest users did not benefit much. Rather, restrictive policies disproportionately affected poorer people and marginalized indigenous groups, especially those living in Terai (Adhikari and Dhungana 2010).

Despite rapid change in forest policies in the Panchayat period, the main concern of marginalized and disadvantaged people was still inadequate access to land. This was a result of historical injustice, which made a large section of the population landless or near landless, especially Dalits and other disadvantaged ethnic groups. In Nepal, about 87% of land ownership comes from inheritance, and so those people whose ancestors did not have land remained without land (CBS et al. 2006). About 7% of households occupy 31% of the land, but the bottom 20% own only about 3% (CBS 2011). Only about 19% of land is under women's name, and most of it is poor-quality land (Adhikari 2009).

After the Ranas were overthrown in 1951, political parties working underground helped various protests of landless and small farmers to access land (Seddon 2018). Some groups, such as the Kamaiyas (bonded labor), were also able to make some covert resistance – even though this was mostly small and symbolic. These collective actions rose only because of the great misery imposed on Kamaiyas (Seddon 2018). For example, in April 1951, the *Kamaiyas* of Bardiya district launched collective action to capture *Khet* and *Khaliyan* (land and barn). Other Kamaiya movements in this period include Srikainda, Dalla, Majhara, Damauli, and Manau, and these all were related to accessing land for farming. These movements were not recognized as legitimate by the state, and thus Kamaiyas suffered severe reprisals.

Democratic Period (1990 to Present)

After 1990, major changes in politics and demography greatly impacted people's environmental concerns and government policies. A major development was legal acts and regulations that aimed at empowering communities in the management of forests. For example, the Forest Act 1993 was instrumental in devolving management to the user committees. This created conditions for rapid expansion in community forestry in Nepal, leading to an increase in forest cover and a reversal of the trend of deforestation, especially in the hills. But the concerns over equitable access and benefits and representation for the poor and marginalized remained. For example, Thapaliya, Jana, and Ghimire (n.d.) examine how Dalits (Chamars) were not given membership to a local community forestry users group initially and how they had to protest for a long time before being granted leadership positions within the community forestry program.

The struggle for land rights continued to be the major form of civil movement in Nepal. There has been some progress especially in bringing about policies for land distribution, tenancy rights, and rights of women in inheriting parental property. Moreover, an increasing number of tenants brought suits for their tenancy rights in various courts (Khadka 2020). This shows that concerned people are taking every opportunity – from collective to legal action – to get justice.

The rapid urbanization of the country has brought in new worries regarding the protection of the urban environment and new struggles for environmental justice. People have grown concerned about urban open space and parks, the dumping of garbage, air pollution, and the health of children. Increasingly, young people are also concerned with climate change and its impact on their lives. For example, Harin Nepal, a youth group established in 2018, regularly organizes movements and protests for issues like climate justice, the rights of people affected by infrastructure, and the like as these young people feel that their future is at stake because of climate injustice and other environmental issues. Harin Nepal was instrumental in organizing a protest to stop the felling of 2.4 million native trees to build a proposed international airport in Nijgadh. In 2022, the legal activism of environmental lawyers led the Supreme Court to stop the construction of the airport in its present form.

Khadka (2022) writes about how difficult it is to organize such movements as activists work voluntarily with a lot of risks; she gives an example of a young person killed in Terai for speaking against the illegal excavation of sand and stone. Khadka also observes that most of these activists fighting to protect marginalized people from hazards or insecurities arising from environmental and development problems and programs are women (60%–70% depending upon the activity[3]).

Two Recent Case Studies

Case Study 1 Kamaiya Freedom Movement

In the western Terai, Kamaiyas (bonded labor), who belong to Tharu ethnic group, have occasionally taken collective actions since the 1950s. However, prior to 1990, their collective action was sporadic. These actions did not develop into full-fledged movements because of the repressive political system of the time. But since 1990, Kamaiyas have used a multipronged approach to press their demands for access to land. That demand was aimed at correcting historical injustices done to Tharus in the process of dispossessing them of their land and enslaving them. This was also meant for livelihood security and for the better care of land, as Tharus were the traditional stewards of the land. They argued that, if given the opportunity, they would care better for the land because their livelihood depends on it.

Starting in the early 1990s, Kamiayas revived their traditional organizations for expressing grievances. They started their protest with a hunger strike to pressure the government to address the loss of their traditionally occupied land. This was called the Kanara movement and led to incremental fulfillment of their demands by the government. But the structural issues like debt bondage, enslavement, and ownership of land were not addressed even though the 1990 Constitution had declared that debt-bondage was unconstitutional. Many NGOs and INGOs and donor agencies helped Kamaiya organizations (particularly BASE – Backward Society Education) in their struggles.

With support from both national and international civil society, the Kamaiya Liberation Movement started on May 1, 2000. It forced the government to declare the emancipation of approximately 200,000 bonded laborers (Fujikura 2001). While there are still implementation problems, it can be argued that the creation of democratic spaces, organization of the affected people rooted in their tradition, and alliance with national and international civil society were instrumental in making this social movement at least an initial success.

Case Study 2 Conservation of Native Trees, Chautaras, and Open Space in Pokhara

Pokhara residents and those in other cities have long been concerned with the loss of open space and of chautaras – stone resting platforms with large native trees. Built with religious symbolism, these chautaras used to dot the city landscape. But in the process of rapid urbanization, these native trees were felled and their key roles in the cultural landscape of the city lost. At the same, time urban development consumed many areas of open space, reducing opportunities for children to play outside the home and adults to exercise.

Adhikari (2004) describes how chautaras were created, the extent of their loss, and their potential use in today's city life. In response to this loss, Pokhara residents formed a civil alliance including people from different walks of life – young environmentalists, religious leaders, and other concerned citizens.

After decades of activism, the movement received a boost when a provincial government was formed in 2017 in accordance with the new decentralized Constitution (2015). The issue aligned with the concerns of the local politicians. Because of this support, a law was unanimously passed in the Kaski provincial parliament regarding the protection of such structures and trees along with other remaining open spaces, and the construction of new parks and recreational grounds.

This case shows three interrelated requisites for the success of a movement – evidence-based research, its communication with the ordinary people concerned with the problem and with decision-making authorities, and an empowered local government. A wide networking with other interrelated movements was also important as alliances with religious groups, youth organizations, human rights, and women's welfare drives created a joint movement.

Conclusion

This study reveals that while the emergence of environmental justice as a concept is relatively new in Nepal, concerns for social justice in environmental matters have a long history in the country. The basic elements of environmental justice movements can be traced back to social movements of several decades ago – whether passive or active. In the past, collective action represented mainly a concern for access to land and other natural resources for the survival of marginalized groups that included women, lower castes, and indigenous peoples. In the repressive Rana and Panchayat regimes prior to 1990, resistance was passive or symbolic, and movements were sporadic and localized. Although mostly passive, these forms of resistance in fact indicated a great intensity of concern.

However, since the rise of democracy in the 1990s more active movements have mobilized access to clean air, parks and recreational grounds, and inclusive urban infrastructures like transportation. Youth and women are highly involved in movements concerned with long-term sustainability and problems affecting family health, especially that of children and the elderly. As such, Nepal positions itself at the forefront of emerging environmental justice movements around the world.

Notes

1 Personal Communication through email with Avash Bhandari, an environmental historian (2023).
2 Personal communication with the first warden who established the park. Also see https://www.aljazeera.com/features/2018/4/29/living-in-the-shadow-of-nepals-rara-national-park.
3 Interestingly, Gaard (2017) states that women comprise an estimated 60%–80% of members in environmental organizations worldwide, and an estimated 90% of members in US environmental justice organizations.

References

Adhikari, J. (2004). A Socio-Ecological Analysis of Loss of Public Properties in an Urban Environment: A Case Study of Pokhara, Nepal. *Contributions to Nepalese Studies*, 31(1), 85–114.

Adhikari, J. (2009). *Land Reform in Nepal. Problems and Prospects*. Kathmandu: ActionAid Nepal and NIDS.

Adhikari, J., & Dhungana, H. (2010). The State and Forest Resources: An Historical Analysis of Policies Affecting Forest Management in the Nepalese Tarai. *Himalaya*, 29(1). https://digitalcommons.macalester.edu/himalaya/vol29/iss1/4.

Auyero, J., & Swistun, D. A. (2009). *Flammable: Environmental Suffering in an Argentine Shantytown*. Oxford and New York: Oxford University Press.

Caplan, L. (2000). *Land and Social Change in East Nepal. A Study of Hindu-Tribal Relations*. Kathmandu: Himal Books.

CBS (2011). *Nepal Living Standards Survey*. Kathmandu: CBS.

CBS, World Bank, DFID, and ADB (2006). *Resiliency Amidst the Conflict. Assessment of Poverty Reduction in Nepal 1995/96 and 2003/04*. Kathmandu: CBS, World Bank, DFID and ADB.

Eckholm, E. (1976). Losing Ground. *Environment: Science and Policy for Sustainable Development*, 18(3), 6–11.

Fisher, R. J. (1994). Indigenous Forest Management in Nepal: Why Common Property Is Not a Problem. In Michael Allen (Ed.) *Anthropology of Nepal: Peoples, Problems and Process* (pp. 64–81). Kathmandu: Mandala Publications.

Fujikura, T. (2001). Emancipation of Kamaiyas: Development, Social Movement, and Youth Activism in Post-Jana Andolan Nepal. *Himalaya*, 21(1). https://digitalcommons.macalester.edu/himalaya/vol21/iss1/14.

Gaard, G. (2017). Feminism and Environmental Justice. In Ryan Holifield, Gordon Walker, and Jayajit Chakraborty (Eds.) *The Routledge Handbook of Environmental Justice* (pp. 74–88). London: Routledge.

Gilmour, D. A., & Fisher, R. J. (1991). *Villagers, Forests, and Foresters: The Philosophy, Process, and Practice of Community Forestry in Nepal*. Kathmandu: Sahayogi Press.

Gurung, R. (1997). *Forest Management by the Gurungs and Bahuns in the Annapurna Range: A Comparative Study*. MA Thesis, Dept. of Sociology and Anthropology, TU., Nepal.

Hamilton, F. B. (1819). *An Account of the Kingdom of Nepal: And of the Territories Annexed to this Dominion by the House of Gorkha*. Edinburgh: Archibald Constable and Company.

Ives, J. D., & Messerli, B. (2003). *The Himalayan Dilemma: Reconciling Development and Conservation*. London: Routledge.

Karki, N. (2022). A Tale of Displacement for Conservation. *The Record Nepal*, February 11, 2022. https://www.recordnepal.com/a-tale-of-displacement-for-conservation.

Khadka, R. K. (2020). History of Land Rights Movement in Nepal. *Online Khabar*, December 26, 2020. https://english.onlinekhabar.com/history-of-land-rights-movement-in-nepal.html, accessed 3 March 2023.

Khadka, S. (2022). Women Leading Environment Justice Movements in Nepal. *Feminist India*, July 6, 2022. https://feminisminindia.com/2022/07/06/women-leading-environment-justice-movements-in-nepal/.

Nesheim, T. (1992). Changes in Forest Management among the Kulunge Rai of East Nepal. *Himalayan Research Bulletin*, XII(1–2), 88.

Paudyal, G., Carr, J., & Munro, P. G. (2022). Community Forestry in Nepal: A Critical Review. *International Forestry Review*, 24(1), 43–58.

Regmi, M. C. (1971). *A Study in Economic History of Nepal 1768–1846*. New Delhi: Munjishree Publishing House.

Regmi, M. C. (1999). *Imperial Gurkha. An Account of Gorkhali Rule in Kumaun*. New Delhi: Adroit Publishers.

Regmi, M. C. (2002). *Nepal: An Historical Miscellany*. New Delhi: Adroit Publishers.

Robertson, T. (2021). Centennial of Hunting Diplomacy in Nepal. *Nepali Times*, December 14, 2021. https://www.nepalitimes.com/here-now/centennial-of-hunting-diplomacy-in-nepal/.

Scott, J. (1985). *Weapons of the Weak: Everyday Forms of Peasant Resistance*. New Haven: Yale University Press.

Seddon, D. (2018). Resistance, Rebellion, Revolt and Revolution: A Historical/Anthropological Consideration of Peasant Movements and Other Forms of Rural Unrest in Nepal. In L. P. Uprety, S. Dhakal, & J. Basnet (Eds.), *Peasant Studies in Nepal* (pp. 25–100). Kathmandu: Vajra Books Inc.

Shah, S. (1991). *Changes and Continuities in an Uprooted Community. A Study of the Rarals in far Mugu District*. MA Thesis, T.U., Nepal.

Sicotte, D. M., & Brulle, R. J. (2017). Social Movements for Environmental Justice through the Lens of Social Movement Theory. In Ryan Holifield, Jayajit Chakraborty and Gordon Walker (Eds.) *The Routledge Handbook of Environmental Justice* (pp. 25–36). London: Routledge.

Thapaliya, A., Jana, S. J., & Ghimire, S. (n.d.). Local Struggle towards Grass Root Democracy: A Case of Terai Dalit Movement for Right to Community Forestry in Eastern Nepal. ODI, CSO Case Study 6. https://cdn.odi.org/media/documents/3619.pdf.

4 Environmental Justice and the Role of Nepalese Judiciary

A Missed Opportunity

Jony Mainaly

Introduction

Although the term 'environmental justice' is often mentioned in judicial decisions in Nepal, there has been limited exploration of its precise meaning and dimensions. This gap in understanding leaves room for interpretation and raises questions about how the court invokes and defines the concept of environmental justice in its decisions. Further exploration and clarification of the meaning and dimensions of justice in the context of environmental justice is essential to ensure a comprehensive understanding of this concept and its implications in Nepal's legal system. This clarity in terms is also crucial to help inform the use of the Supreme Court's interpretation by environmental justice organizations and advocates in Nepal.

Nepal's judiciary has played an instrumental role in developing and advancing environmental jurisprudence for decades. This has been particularly important because Nepal's Constitution did not have explicit provisions on the right to a clean and healthy environment in the 1990 constitution. However, the judiciary has played an instrumental role in expanding the interpretation of the right to life to include the right to a clean and healthy environment through its judicial pronouncements (Surya Prasad Dhungel vs. Godavari Marble Industry and others, 1996, p. 169; Yogi Naraharinath vs. Office of Prime Minister and Others. NKP, 1997. No 1. Decision No. 6127)

With several environmental protections related cases, the Supreme Court of Nepal has played a paramount role in fulfilling both constitutional and legal vacuum. With the help of its interpretation, the Court can shape legal provisions and its practice.

In this paper, I examine the concept of environmental justice in Nepal through an analysis of four selected Supreme Court decisions. The cases chosen are related to hydropower development, the marble mining industry, brick kiln industries causing air pollution, and encroachment and infrastructure development around Phewa lake. These cases involve critical natural resources such as water, forests, and biodiversity.

Through these cases, it is clear that Nepal's judiciary uses the term environmental justice very narrowly to mean environmental protection. In doing so, I argue that Nepal's judiciary has missed an opportunity to clearly invoke the multiple important dimensions of environmental justice. The idea of environmental justice originating

DOI: 10.4324/9781003371175-5
This chapter has been made available under a CC-BY-NC-ND 4.0 license.

from the United States conceives of environmental justice through an intersectional lens wherein diverse socio-economic groups do not face unfair environmental burdens, appears conspicuously absent in Nepal's judicial pronouncements.

The Three Dimensions of Environmental Justice

In the United States, the meaning of environmental justice was shaped by the 1980s environmental justice movement. However, the meaning of environmental justice has evolved in different regions and countries over time (Schlosberg, 2007; Beretta, 2012). Some scholars have focused only on distributional issues (Figueroa & Mills, 2001) while others emphasize procedural and recognition dimensions. Even within the US Environmental Protection Agency in the 1980s and 1990s, different administrative levels and programs have created their definitions of environmental justice (Rhodes, 2003). However, overall, the environmental justice movement has stressed that the share of the burden and benefits of environmental resources have often been uneven based on race. It highlighted that environmental injustice is linked to socio-economic injustice.

Belbase and Thapa conducted one of the first empirical studies on environmental justice in Nepal, identifying critical elements of environmental justice as 'equitable access to and use of natural resources, participation in decision-making and management, distribution of benefits without discrimination, and exercise of rights to natural resources' (Belbase & Thapa, 2007, p. 95). They provide insights into the dimensions of environmental justice in rural Nepal, which aligns with discourses on environmental justice from other parts of the world. Nevertheless, examination of the academic literature, legal texts, and judicial pronouncements shows that few scholars have deployed the key ideas of environmental justice in Nepal.

Although there are different dimensions of environmental justice, the literature predominantly focuses on distributive environmental justice and procedural. For example, Peter S. Wenz defines environmental justice exclusively in distributive terms (Figueroa & Mills, 2001). On the other hand, recognition justice is 'arguably the most neglected and under theorized' in environmental justice realm (Coolsaet & Néron, 2020, p. 2).

Distributive

The distributive justice principle advocates for the fair distribution of environmental benefits and burdens within society (Lazarus, 1993), ensuring equitable allocation of environmental resources, such as clean air, water, forest while mitigating the disproportionate impact of pollution and environmental degradation on vulnerable individuals and communities.

Using an example of a landfill site, Kaswan (2003) distinguishes between two claims under the distributive justice claim—the narrow and the broader one. The narrow claim focuses on the disproportionate burden borne by the local community where the landfill site is located, while the broader claim addresses the systemic inequities leading to the selection of marginalized communities as landfill sites.

The central idea of the distributive justice principle lies in the outcome of the resource allocation rather than the fairness of the distribution process itself (Rawls, 1999, p. 24).

Procedural

The principle of procedural justice prioritizes the fairness of decision-making processes, asserting that individuals should not be subject to decisions that detrimentally affect them without their participation. It hinges on the means-to-end argument, positing that fair decision-making processes generally lead to fair outcomes. H.L.A. Hart, a prominent legal philosopher, argues that decisions made without considering the interests of all sections of the community are unjust and open to criticism (Hart, 1961, p. 163).

Central to procedural justice is the promotion of 'open, informed and inclusive decision-making processes,' (Gonzalez, 2012, p. 4) particularly during any development project's Environmental Impact Assessment (EIA) process. Nepal's environmental legislation mandates public involvement through various means such as holding meetings, ensuring access to information; providing comment periods; and holding public hearings (Environment Protection Act (2019) (Nep); Environment Protection Regulation (2020) (Nep).

Although an essential precondition for procedural justice, participation procedures do not guarantee that the final decision embodies the public voice. In cases of participation lacking substantive and meaningful engagement, procedural justice remains unrealized.

Recognition

The scholars who emphasize justice as recognition, such as Nancy Fraser, assert that distribution issues are essential but incomplete to understand justice (Young, 1990). She contends that justice entails not only determining fair distribution but also examining the factors that contribute to unfair distributions (Fraser, 1997). Although distributional justice answers 'who gets what,' recognition justice answers 'why does one get what one gets.' In several of his articles, Schlosberg emphasizes that environmental justice has 'never been about equity alone' (Schlosberg, 2007, pp. 4–5; Schlosberg & Collins, 2014, p. 361). It questions why already vulnerable and marginalized people and communities face distributional inequity (Schlosberg & Collins, 2014). To achieve environmental justice, policies and programs must acknowledge local concerns, promote economic empowerment, and demonstrate respect for ethnic and cultural diversity (Pulido, 1996).

Role of Courts and Public Interest Litigation in Environmental Justice

According to Outka, 2005, p. 241 environmental injustice is a 'widespread and complicated' problem that the judiciary alone is not equipped to solve. However, judiciaries have a significant role to play.

With relaxed standards allowing virtually anyone to approach the court with issues plaguing the public, the Public Interest Litigation (PIL) system in Nepal has allowed the Court to adjudicate and pronounce decisions with wide implications. Article 133 of the Constitution of Nepal entrusts every Nepali citizen with the right to approach the Supreme Court directly 'for the settlement of any constitutional or legal questions involved in any dispute of public interest or concern.' This is called the extraordinary jurisdiction of the Supreme Court of Nepal.

Under PIL regime, Nepal's Supreme Court has been instrumental in pronouncing environmental standards. Nepal's apex court has given decisions on 'clean vehicles,' the appointment of environmental inspectors, the closure of polluting industries, and climate change, among others.

Through this rather empowering tool for any publicly concerned individual to reach out to the court of law, the court gets room to assess procedural rules and the rule of law at large. Any individual or organization concerned about environmental issues can bring lawsuits against polluting industries or government agencies that have failed to enforce environmental laws and regulations. By ensuring PIL rights to any individuals and following a liberal interpretation of the concept, Nepal's Constitution has made environmental justice remedies more accessible.

Hon. Dr Anand Mohan Bhattarai, Justice of the Supreme Court of Nepal, has shed light on the role of judges in South Asia, departing from the traditional role of norm interpretation and embracing a norm-making function (Bhattarai, 2012). Through the Court's approach to interpret traditionally non-justiciable rights, environmental rights found their place in the fundamental rights section of the Constitution. With such a norm-making function of the Supreme Court, the right to a clean and healthy environment found its place in the fundamental rights chapter of the 2007 Interim Constitution.

In the environment-related PILs in Nepal, lawyers are the petitioners who bring these cases to the court on behalf of the general public. The case could regard a polluting river in Kathmandu or polluting industries in Lumbini almost 300 KM from Kathmandu. At times, the Court has warned petitioners to avoid 'Publicity Interest Litigation' in the name of Public Interest Litigation (Ashok Kumar Pandey v. State of W.B., 2004 (3) SCC).

The following section analyses four environmental cases brought before the Supreme Court of Nepal under Public Interest Litigation.

Ramchandra Chataut vs. Government of Nepal, Office of Prime Minister and Council of Ministers et al., NKP 2066 [2009], DN. 8059

In this case, the Supreme Court of Nepal emphasized environmental justice in its decision involving hydropower project development.

The petitioner filed against the government's agreement with an international company (Snowy Mountains Engineering Corporation Limited) to develop a 750 MW hydropower project in Nepal far western region. The petitioner raised concerns about environmental impacts, local people's participation in the decision-making process, their right to other beneficial use (such as irrigation) of discharged water after hydropower generation, and the displacement of households from the project sites. Overall, the petitioners noted that there are no arrangements for access to benefits for the local people who have been the custodians of the resource for generations and who face disproportionate environmental risks.

The court emphasized the need for sustainable use of natural resources, fair treatment of local communities, information sharing, environmental mitigation measures, compensation for harm, and relocation of displaced communities. The court also highlighted the importance of informed consent, language accessibility, and continuous information flow to local communities. The technical nature of the information and the lack of simplicity in communication often result in local communities being unable to fully utilize the EIA process. This 'creates a gap in understanding between the project and local communities, which could lead to feelings of deception,' the Court wrote.

The Court established that a development project involving the use of natural resources cannot be considered environmentally just if local communities are not engaged, their consent is not sought, or if their complaints and difficulties arising from the project are ignored.

The displacement of people is a delicate matter that significantly affects a community's traditional lifestyle and livelihood. Relocation decisions must be made with the informed consent of the communities and should not be imposed upon them. Local people are the custodians of natural resources, but the state and private entities reap their benefits. This is injustice per se and discourages local communities from resource conservation.

This case has implications for environmental justice. The court rightly states that it is a state of environmental injustice if the project degrades the quality of life of local communities.

However, neither the petitioners nor the court stresses the demographic composition of the affected population. Apart from being rural and local, their demographic identity is not mentioned. Far western region is one of Nepal's most underprivileged regions and its rural regions are most certainly economically disadvantaged. More discussion on social and ethnic identity would be helpful in making further judgments about environmental justice concerns.

The court issued a *mandamus* order in the name of the Ministry of Water Resources to formulate an Environment Protection and Remedy Committee. However, it did not quash the agreement between the government and the company for failing to observe all the abovementioned environmental justice relevant concerns. The decision appears to be more rhetorical than real.

Prakashmani Sharma vs. Government of Nepal, Secretariat of the Council of Ministers et.al, NKP, 2073 [2016], DN. 9575

The court's decision in this case asserts that the notion of environmental justice is intrinsically linked to the principles of 'sustainable development and equitable benefit sharing,' as delineated in the directive principles and policies of the state outlined in the Constitution. However, the decision falls short of explicitly connecting this assertion with the facts of the case at hand.

A writ petition was filed against Godavari Marble Industry, a marble quarrying factory in the Kathmandu Valley. The petitioners claimed that the industry, situated in a forest area that is rich in bio-diversity and ecological significance, had caused environmental damage, including deforestation, bio-diversity loss, pollution of rivers, streams and water bodies, and noise pollution. The Supreme Court ordered the factory to close its doors.

Article 35(5) of the Interim Constitution 2007 provides for state's responsibility towards 'protection of forest, vegetation and biodiversity, their sustainable use and equitable distribution of benefits derived from them.'

The decision sheds light more on inter-generational equity where it states 'it is important to have futuristic approach when we are discussing environmental justice.' The court without reflection on the factual basis, provides a definition of environmental justice from the Environmental Protection Agency of the United States. It neither shows this definition's relevance to the facts or to this ruling. The use of the environmental justice principle does not reveal the judicial wisdom that would allow future justices to reason their way to the best outcome.

Khagenra Subedi et.al vs. Government of Nepal, Office of Prime Minister and Council of Ministers et.al, NKP 2075[2018], No.9, DN: 10086

Although mostly in rhetoric again, the Supreme Court of Nepal in this case also discussed intra-generation equity in addition to inter-generational equity.

This case is about the protection of Phewa Lake against encroachment and degradation. Situated in a tourist hotspot of Pokhara, the lake provides multidimensional values and benefits. It is rich in natural, cultural, and religious resources. Construction of hotels and lodges has encroached on the lake, and lake pollution has increased due to nearby residents. The development has undermined irrigation, aquatic plants and animals, and the tourism and environmental balance of the area.

The court handed down the orders of *mandamus* to ensure protection, both against encroachment and for maintaining ecological balance.

The court emphasized inter-generational and intra-generational equity in relation to natural, cultural, and environmental resources, as well as the rights of local communities. It emphasizes the need to ensure the availability of these resources to future generations while also addressing the needs and rights of different population groups within the current generation.

The case highlights how the court assessed the obligation of the government in ensuring inter-generational equity, sustainable use, and equitable benefit sharing of environmental resources, as outlined in the directive principles and state policy chapter of the Constitution (2015, 51(g)). The court specifically examined the provision related to providing priority and preferential rights to local communities in managing and benefiting from these resources as provided in the Constitution.

Prakashmani Sharma vs. Cabinet Secretariat, HMG, Singh Durbar, et al., NKP 2062 [2006], DN. 7581

In questions to be decided, the court in this case looked at the important question of 'whether or not environmental pollution fall under the scope of environmental justice.' While an extremely important question, the court again resorted to rhetoric instead of establishing a norm that environmental pollution falls under the scope of environmental justice.

The petitioner filed suit against the unauthorized operation of brick kiln industries in Kathmandu Valley. The unregulated operation of such industries is largely responsible for air pollution in the Kathmandu Valley, thereby invalidating citizens' fundamental right to live in a clean and healthy environment.

The court issued an order to assess the number of brick kiln industries, the amount of pollution emitted by those industries, their adoption of emission reduction technology, the potential impacts on infrastructure development and residential construction due to the closure of such industries, and the location of schools and dense human settlement.

The rights to occupation, employment, business, and industries are fundamental in the Constitution. However, it is not permissible to operate a factory at the expense of public health.

This case also took environmental justice as an important question, asking whether or not environmental pollution falls under the scope of environmental justice. In the final decision, the Court addresses the distributive dimension of environmental justice in assessing the benefits that the industries bring to industrialists, workers, and general consumers compared with the adverse environmental pollution they cause. Based on the principle of private interest must yield to public interest, the pollutant industries must be brought within higher scrutiny.

This case is important as the decision came at a time when the right to clean and healthy environment was not explicitly provided as a fundamental right but the right to occupation, and employment was. The decision states that environmental pollution endangers the citizen's right to live in a pollution-free environment and right to life.

Analysis and Discussion

In discussing cases related to environmental justice in Nepal, several themes emerge.

First, the judicial rulings encompass three major dimensions of environmental justice—distributive, procedural, and recognition—although detailed discussion on these dimensions is rare. For example, the case of brick kiln industries, hydropower, and Phewa Lake delve into discussion of distributive justice without mentioning the term. In terms of recognition dimensions, the Court discusses the enforcement of fundamental rights and the provisions of laws. None of the cases, however, discuss the specific socio-economic identities of the people and communities involved.

In all of the four cases, the distribution of harms and benefits is discussed from the perspective of local people's harms and benefits. The court does not, however, identify the local population's economic, social, ethnic and cultural characteristics. For example, Ram Chandra Chataut Case focuses on the role of local communities in resource conservation and the potential negative impacts that they are at risk of bearing from the dam. Similarly, in the Godavari Marble, Phewa Lake, and brick kiln cases, the Court assesses the environmental harms the local community, schools, local hospitals, and resorts face due to industrial pollution and encroachment. The distributional analysis is based on locality, not economic and/or social identities. The Court generally ignores specific demographic disparities, severely undercutting their approach to environmental justice.

Relatedly, within distributive justice, most of the related cases focus on inter-generational equity and justice more than intra-generational equity. All the cases discuss in detail the importance of conservation for future generations. This focus may be because inter-generational equity concerns are provided in the Constitution Article 51 and also a part of international environmental laws and policies. However, the Court mostly overlooks the question of sustainable development, which is mentioned as a balance between the environment and development in the Constitution's fundamental rights section, thereby missing a chance to focus on intra-generational equity.

Second, all the Supreme Court cases analyzed focus on ecological justice, not environmental justice. The term ecological justice differs from environmental justice. According to Schlosberg, ecological justice relates to justice between humans and nature (conservation focus) and environmental justice relates to justice between human beings on environmental concerns (distribution focus) (Schlosberg, 2007). All the analyzed cases are ecological justice focused which emphasizes the conservation of the environment.

Third, these decisions on environmental justice often involve law enforcement. This includes holding entities accountable for violations of environmental laws and regulations, sometimes imposing penalties. It also recognized the role of government in enforcing environmental laws and regulations. In doing so, the Court has ordered the government to take action to protect the environment and has nullified decisions that are potentially harmful to the environment and public health. But none of the cases focus on the potential harmful activities to disadvantaged populations.

Fourth, although the Court in many of the cases discussed took up environmental justice questions, it does not discuss them in a detailed manner, elucidating the different dimensions of environmental justice. The court uses the term environmental justice in all of these cases but does not make its analysis detailed regarding the case facts. Also, it discusses environmental justice concerns in *obiter dicta* but does use environmental justice for the decision's basis *(ratio decidendi*). *Obiter dicta* is a judge's observation or comment when deciding a case. *Obiter dictum* can be used as a persuasive authority for future litigation but it is not legally binding as *ratio decidendi* is. This may be because the concept of environmental justice per se is not provided directly in the Constitution or the law. However, as mentioned above, with PIL jurisdiction, the Court may go beyond the realm of constitutional and legal provision, and can invoke the meaning of environmental justice.

The Court's use of environmental justice demonstrates it is inclined to norm interpretation alone, not norm-making function. The heightened role the Court has played in advancing environmental jurisprudence is limited when it comes to environmental justice. The environmental jurisprudence Nepal's judiciary has advanced includes environmental protection, intergenerational theory, and public trust doctrine that is termed as ecological justice but not much in environmental justice. This has force as a principle, but because environmental justice does not form a corpus of constitutional/legal provisions, the decision directly does not mention the same. This means environmental justice remains merely as rhetoric and not a norm.

Conclusion

The concept of environmental justice has been frequently mentioned in judicial decisions in Nepal but has not been thoroughly explored or defined. This lack of clarity raises questions about the Court's understanding and application of the concept. Environmental justice encompasses various dimensions, including distributive, procedural, and recognition, and it has implications for diverse individuals and communities. Most importantly, the Court has conceptualized local communities as monolithic, not diverse. Because of this problem, Nepal's judiciary has missed out on the opportunity to advance a norm for environmental justice.

Looking into the cases in the Supreme Court of Nepal on environmental matters raises a disturbing question that goes unanswered in the court's judgments—Are marginalized communities and individuals burdened with more than their fair share of environmental risks while enjoying fewer of the benefits or environmental laws than others? It is also troubling that very few environmental justice cases are brought to the court by the affected parties. Instead, mainly Kathmandu-based lawyers based are filing these petitions. This suggests that the general public may not be well-informed or willing to bring cases to the court. This raises important questions about access to justice. Further research and examinations are necessary to understand why this is the case. Such research on the relationship between the Court and Nepal's social movements will be invaluable.

Because most of the PILs brought to the Supreme Court of Nepal are introduced from an environmental protection perspective, environmental justice for the many Nepali populations dependent on environmental resources is often overlooked. In an adversarial justice system, the court only hands down decisions based on the claims brought by the petitioners. However, aware of the extent of the problem, the Supreme Court has propounded several decisions by exercising judicial activism under the PIL regime. Unfortunately, the judiciary has applied only a narrow environmental protection approach not a broader environmental justice approach. This has therefore been largely a missed opportunity.

References

Ashok Kumar Pandey v. State of W.B., 2004 (3) SCC 349.

Bhattarai, A. (2012). Access of the Poor to Justice: The Trails and Tribulations of ESC Rights Adjudication in South Asia, NJA L.J. 1, 29 (special issue).

Belbase, N., & Thapa, L. (2007). Environmental Justice and Rural Communities: Studies from India and Nepal. IUCN. Available at https://portals.iucn.org/library/sites/library/files/documents/2007-030.pdf

Beretta, I. (2012). Some Highlights on the Concept of Environmental Justice and its Use. OpenEdition Journals. Available at https://journals.openedition.org/eces/1135

Coolsaet, B., & Néron, P. Y. (2020). Recognition and Environmental Justice. In B. Coolsaet (Ed.) *Environmental Justice* (pp. 52–63). New York: Routledge.

Figueroa, R., & Mills, C. (2001). Environmental Justice. In D. Jamieson (Ed.), *A Companion to Environmental Philosophy* (pp. 426–438). Blackwell Publishers.

Fraser, N. (1997). *Justice Interrupts: Critical Reflections on the 'Postsocialist' Condition.* New York: Routledge.

Gonzalez, C. (2012). Environmental Justice and International Environmental Law. In E. J. Techera, J. Lindley, K. N. Scott, & A. Telesetskty (Eds.), *Routledge Handbook of International Environmental Law* (pp. 77–98). Routledge.

Government of Nepal. (2019). Environment Protection Act, 2019.

Government of Nepal. (2020). Environment Protection Regulation, 2020.

Hart, H. L. A. (1961). *The Concept of Law*. 2nd Edition. London: Oxford University Press.

Kaswan, A. (2003). Distributive Justice and the Environment, 81 N.C.L.Rev.1031.

Khagenra Subedi et al vs. Government of Nepal, Office of Prime Minister and Council of Ministers et al, NKP 2075[2018], No.9, DN: 10086.

Lazarus, R. (1993). Pursuing Environmental Justice: The Distributional Effects of Environmental Protection. *Northwestern University Law Review*, 87(3): 787–857.

Outka, U. (2005). Environmental Injustice and the Problem of the Law. *Maine Law Review*, 57, 209.

Prakashmani Sharma vs. Cabinet Secretariat, HMG, Singh Durbar, et al., NKP 2062 [2006], DN. 7581.

Prakashmani Sharma v Godavari Marble, et al., NKP 2016. No 4. Decision No. 9575.

Prakashmani Sharma vs. Government of Nepal, Secretariat of the Council of Ministers et al., NKP, 2073[2016], DN. 9575.

Pulido, L. (1996). *Environmentalism and Economic Justice: Two Chicano Struggles in the Southwest*. Tucson: The University of Arizona Press.

Ramchandra Chataut vs. Government of Nepal, Office of Prime Minister and Council of Ministers and others, NKP 2066, DN.8059. Available at https://nkp.gov.np/full_detail/4003

Rawls, J. (1999). *A Theory of Justice*. Rev. ed. Cambridge: Harvard University Press.

Rhodes, E. (2003). *Environmental Justice in America: A New Paradigm*. Bloomington: Indiana University Press.

Schlosberg, D. (2007). *Defining Environmental Justice: Theories, Movements, and Nature*. Oxford: Oxford University Press.

Schlosberg, D., & Collins, L. (2014). From Environmental to Climate Justice: Climate Change and the Discourse of Environmental Justice. *WIREs Climate Change*. Doi: 10.1002/wcc.275

Surya Prasad Dhungel vs. Godavari Marble Industry et al., Nep. 1996, p. 169.

Yogi Naraharinath vs. Office of Prime Minister et al., NKP, 1997. No 1. Decision No. 6127.

Young, I. (1990). *Justice and Politics of Difference*. Princeton, NJ: Princeton University Press.

Part 2

Land, Forests, and Agriculture

5 Environmental Injustice in Confronting Gendered Access to Land in Nepal

Joint Land Ownership as a Promising Practice

Srijana Baral, Kalpana Karki and Kanchan Lama

Introduction

Background

Gender disparities in land access remain significant around the globe as land is considered a significant indicator of wealth, power, political, and social security (IOM, 2016). Women's ability to own and control land is essential for their empowerment (Agrawal, 1994b). Access and control over land is a major shaper of social relations. Women holding land ownership in practice is a requisite for gender equality (Agrawal, 1994a). And even if they have land ownership, the quality of land and lack of control over what they produce is often minimal (Akter et al., 2017). Women's right to access and control over land strongly influences their overall living conditions, economic security, and physical safety and has a profound effect on gender relations (IOM, 2016).

Globally women own less than 20% of the land despite campaigns like "Stand for her land" (Villa, 2017). The situation is slightly better in Nepal with 23.8% of land ownership by women (GoN, 2021). Local social norms and everyday gendered politics marginalize women from access to land in Nepal and elsewhere. As a consequence, Nepal's skewed distribution of land undermines the socioeconomic and political standing of women, especially on household-level decisions.

Conversely, the redistribution of land ownership and increased recognition of women's access to land tenure has been linked with their social and economic resilience and greater say in decision-making at the household level (Keera, 2007). Access to land resources is found to improve the well-being of children and family as women can have better options and more decision-making power about nutrition (García-Morán & Yates, 2022). The importance of women's access to land resources was signified in the 62nd session of the Commission on the Status of Women of the United Nations, which warned that unless gender equality is extended to land rights and ownership, the calls for greater equality for women in the 2030 Agenda for Sustainable Development would become an impossible vision (WOM, 2018).

DOI: 10.4324/9781003371175-7
This chapter has been made available under a CC-BY-NC-ND 4.0 license.

However, in addition, justice for women needs to move beyond the distribution of environmental goods (e.g. land distribution) to also consider full recognition of their access to land and their increasing participation in decision-making (Chaudhary et al., 2018; Fraser, 1997) when it comes to land and other properties. Land ownership and associated rights together are considered an important component of environmental justice and gender equality.

In this chapter, we use an environmental justice lens to explore how the joint land ownership (JLO) program in Nepal can promote gender equality. Environmental justice provides a lens to focus on fair treatment of all with respect to the development, implementation, and enforcement of environmental laws, regulations, and policies (Schlosberg, 2004; Fraser, 1997; Young, 1990). Injustices occur from a lack of recognition or misrecognition of issues related to social categories like caste, class, gender and culture. The land rights of women can be seen as elements in the social basis of self-respect, which Rawls (1971) defines as the primary right. Barring women from owning land is a violation of women's human rights (WOM, 2018).

JLO is one important strategy to achieve gender and environmental justice. It allows both men and women to officially register land in their name. This is a great improvement because in Nepal, as in many other countries, there is often gender-based discrimination in land ownership. An environmental justice approach helps ensure that women gain access to the resources they need for their health and well-being.

Thus in this chapter, we showcase the challenges and the best practices of JLO to address gendered injustices. We will establish the usefulness of equality in land rights for both women and men as a way to achieve socially and environmentally just decision-making processes over productive assets.

Social Movements for Joint Land Ownership

Nepal's modern land rights movement started in an organized way in 2005. The tenants who used to plow the land of the landlords pushed to obtain "tillers' rights." Many were eventually granted these rights. However, none of them were women. Since 2009, the Community Self Reliance Centre (CSRC) and Nepal Land Rights Forum (NLRF) have worked to expand women's land rights in Nepal. This work began with policy and advocacy efforts, including an 11-day march in 2010 by 337 land disadvantaged women from 42 districts. In March 2011, a national movement for Timely Constitution Writing and Land Reform took place, where more than 1,000 poor women and 100 men from 50 districts came to Kathmandu to make demands of the government and political authorities. The Government of Nepal made a decision in favor of "Joint Land Ownership" following mobilization by social movements, several weeks' demonstrations, and dialogues with policymakers (CSRC, 2012).

The joint land certificate seems to be just a piece of paper but along with it comes power and equality (CSRC, 2012). With this provision, women's ownership over land was recognized and attempts were made to redistribute land ownership to women. So far, CSRC and the Land Rights Forum have facilitated registration by 10,364 couples of 2,825 hectares of land as "joint land ownership" in what is a historic achievement.

Policy Provisions Favoring Women's Land Ownership in Nepal

Nepal has ratified a number of international treaties and instruments that focus on women's rights regarding land and agrarian reform, including the Universal Declaration of Human Rights, the Convention on Political Rights of Women, the Convention on the Elimination of All Forms of Discrimination Against Women (CEDAW), the International Covenant on Economic, Social and Cultural Rights, the International Covenant on Civil and Political Rights and the United Nations Declaration on the Rights of Indigenous Peoples (UN Women, 2020).

In line with these international treaties and instruments, Nepal has taken a number of positive steps to address disparities in relation to land and property, including the 11th Amendment of the Muluki Ain. Similarly, the Constitution of Nepal and the Financial Bill 2015 also addressed women's equality in property rights. The Constitution of Nepal guarantees the right to own property as a fundamental right; every citizen has the right to "acquire, enjoy, own and sell, reap professional gains, and otherwise utilize or dispose property, subject to the existing laws." In addition, in order to encourage women's land ownership, the Financial Bill 2015 provides exemptions to women on the land registration charge. A single woman with a deceased husband receives a 35% tax exemption in land registration. After the new civil code in 2018, daughters after getting married are entitled to keep their share of their parents' property. After a divorce, the woman is also entitled to their husband's share of the property, if the divorce is the husband's fault (Koirala, 2022). The government charges only Nrs 100 if the husband and wife seek JLO.

Nepal's Constitution Article 38 which supports the rights of women that include rights to lineage, right to safe maternity and reproduction, and equal rights in family matters and property has been a major factor in women's land ownership gains (GoN, 2015). As a result of this and other policies the proportion of households in which women owned land has increased from 10.8% in 2001 to 19.7% in 2011 to 23.8% in 2021.

However, women's control over land in Nepal has not proceeded as far and fast as hoped. Despite the fact that women in Nepal contribute more than 80% of agricultural labor, women's land ownership remains minimal (IOM, 2016). The limited land rights by women and girls indicate that the vast majority remain victims of discrimination and marginalization within a patriarchal society. Without resources such as land, women have limited say in household decision-making, and no recourse to assets during crises (UN Women, 2020). Usually, men are included for consultation about land management matters, as they are regarded as the household heads and the landowners even in abstentia. The lack of recognition of women's rights and contributions creates unequal power relations between men and women. Women's limited economic and political agency often leads to severe poverty (Koirala, 2022) and health and well-being problems (e.g., malnutrition, various diseases, and mental health conditions) (Keera, 2007; Tamang, et al., 2014).

This study seeks to describe the challenges facing women in obtaining procedural, distributive and recognition justice and benefits of JLO in their struggle.

Methodology

Study Area

The study was conducted in Mahottari, Dang, and Rasuwa districts in the Terai, Midhill, and Himalayan regions respectively. Participants were selected based on their involvement in the past land rights movements. Of the 61 individuals surveyed, 47 (25 Female, 22 Male) informants have JLO. Fourteen women deprived of JLO were interviewed. Among the 61 respondents, 26 of them belonged to Brahmin, Chettri, and Thakur (BCT) community while 35 of them belonged to ethnic communities (Magar, Tamang, and Tharu).

Methods

Family Couple's Survey

Family couples both holding joint land ownership and with male-only ownership were interviewed using a structured interview protocol. In total, 24 JLO couples (23 married couples and one team of son and mother) and 14 Non-JLO couples were interviewed.

Focus Group Discussion

Twelve focus group discussions were carried out in the districts to discuss the constraining factors after receiving JLO faced by different caste and ethnic groups, i.e. separate discussions with BCT group and ethnic groups were also carried out. In addition, a discussion was held between men and women having joint land ownership (4), men and women not having joint land ownership (4), landless women (4), and squatters' groups (4). Each focus group discussion engaged four to six individuals.

Results

Existing Gender-Based Injustice in Land Ownership, Its Causes and Consequences

Various gender-based injustices were found to exist among women lacking land ownership. We saw evidence of the traditional and patriarchal mindset that hinders full-fledged implementation of the policies and plans. Although women have rights over their ancestral property either as a daughter or wife, in practice, they cannot compel their parents to give them their share of property while their parents are alive (Koirala, 2022). Further, male members hesitated to share their land ownership with their spouse or women members of the family. In addition, at both the local and district levels, there is a lack of budget allocation for programs to secure women's land rights. Consequently, many women have not been able to take advantage of the legal provisions established for them.

Male-Dominated Land Ownership and Patriarchal Mindset

The inheritance of property has traditionally been patriarchal, in which the ancestral property is passed through the male line. As a result, there remains a wide gap between what is provided for in law and what is actually practiced. In most communities, there is strong male resistance to women's land ownership due to patriarchal values and norms against women's rights to family property. In the case study sites, men discouraged women from JLO by saying, "There is no need to make JLO since their land size is too small." One of the men from a non-JLO participant group shared, "We are consulting women in household discussions and family matters, so why should we share our land certificate with them?" As the property is inherited by men, a man's role is significant in the property rights transfer, especially with the land. In some cases, women (especially the daughter-in-law) can even lose control over the dowry they received from their parents. Some women expressed a fatalistic perspective. One of the female respondents said, "We need to support our family and the decision that best fits the family interest should be my interest, though I might have to compromise and feel bad, after a while I forget it and life goes on."

Exclusion from Accessing Financial Resources

Due to limited land ownership, women often lack access to banking and credit facilities. In several households, women wanted to expand the farm or switch to commercial farming and utilize technology. But in several cases, women were refused a loan because they could not provide any collateral security. This has discouraged women from achieving financial sustainability and set them up for loans from other institutions and even moneylenders with high-interest rates. Consequently, these measures have trapped women in the vicious cycle of financial, social, and political poverty.

Social Stigma

A woman may have the legal right to land and property, and yet the right may not be recognized as socially legitimate due to prevailing social norms and attitudes. In the three study districts, patriarchy and a conservative social structure persist, reducing women's mobility and status. There are certain norms against women owning land. Social norms expressed in our study areas positioned men as the household heads, that it is his property as he inherited it and he thus has the right to hold the ownership. A Chaudhary woman respondent shared that

> There is no system of land division (property distribution) between men and women, so if we think about sharing the land ownership, what will others say? We will be the talk of the town as people think that we don't trust our husband!.

In addition, there is a belief that women cannot handle property and financial responsibilities. "If she cannot handle the responsibilities then there is no question of sharing the land ownership with her," said a male respondent. Besides, the male members think that women with land ownership might elope and they would lose their property shared with her. This social stigma curtails women's rights and hinders their fight against discrimination.

Unwelcoming New Spaces for Women

The process of land registration and transfer is often time-consuming, involving high administrative costs and requiring many different documents. Several middlemen such as legal assistants/deed writers (*lekhandaas*), and paralegal agents charging fees are part and parcel of this registration process. For women, these are unfriendly and even unsafe spaces. Some women reported facing harassment when they engaged in land-related transactions, including buying and selling. Women who visit the land offices recount that they were asked personal questions about their family, husbands, and children, which men did not have to answer. Hence, women prefer to stay away from this space.

Joint Land Ownership as a Means to Correct Injustice in Gendered Land Ownership

Land rights campaigners such as those from CSRC and other organizations encourage families for JLO registration by explaining the benefits for women, men, and the family overall. There are three CSRC land rights campaigners and twelve *Lekhandas* at the district land registration office supporting this campaign. They accompany the women to the land registration office where they could register JLO without any hassle. In some cases, the women who have already registered JLO act as advocates and encourage their neighbors to obtain JLO. Be it husband and wife or a pair of mother and son -- all benefit from the feeling of ownership (Figure 5.1).

Redistribution of Land Ownership for Gender Equality

JLO recognizes the women as the landowners and thereby enhances women's social status. Furthermore, the Government of Nepal has introduced a tax subsidy provision to increase women's access to land ownership. Land ownership has offered women economic security for their old age and for younger women, an option of independence. Some of the JLO participants said that having ownership over a piece of land gave them a place to call their own; otherwise, they would be living at their father's

Figure 5.1 Farmers of Rasuwa district after receiving Joint Land Ownership entitlements.
Photo credit: Community Self-Reliance Center.

home before marriage and at their husband's home after marriage. "Now we have our own land," shared one of the women respondents proudly. With JLO, some women can enjoy more decision-making power about their land in cases where their husband has migrated for work or abandoned them or died.

> The story of Kanchhi Maya and Toyaman Jimba illustrates how crucial JLO is to obtain gender equality. Kanchhi and Toyaman live in Malta 8, Lalitpur and were married 35 years ago. They have two sons and one daughter, already married. They worked hard to raise their three children and even sold a part of their land to get them married. But as they approached old age, their son started misbehaving with Kanchhi and threatened to remove her from the house. In the meanwhile, Toyaman was worried about what would happen to Kanchhi after his death. Who will look after her? One day Toyaman got an opportunity to participate in an interaction program organized by the District Land Rights Forum and came to know about JLO. He instantly transferred his sole ownership into JLO. He remarked, *"Now I can take my last breath in peace, since my wife will be safe during her aging, as she holds the land title."* Kanchhi Maya said, *"Now I also hold the right to the land I worked all my life on. I can also decide about the buying or selling of the land!"*

Enhanced Recognition of Women as Owners

By having JLO certificates, the women's social status has been dignified and secured. They are more motivated to work in "their" fields now. Once they obtain ownership, they also share equal responsibility for making decisions on the use, management, and sale of the property with their husbands. Outsiders also look at them with respect as one respondent shared, "In women's informal talking, other women consider me as privileged because of my land holding status," a Dalit woman from Mahottari. Some of them have started adding on to their houses by taking a joint loan using the JLO status. Others have even participated as role models in JLO campaigns, going door-to-door to encourage JLO registration. A Chaudhary woman from Dang shared how she was able to obtain JLO after her neighbor inspired her,

> I have registered JLO with my son as my husband passed away a few years back. I was motivated to do so by my neighbor. I didn't know about these land provisions until she told me about it. I feel honored, safe and happy. I am also encouraging others to go for this provision.

Another woman from the same community added, "I have a feeling of security, that my land won't be taken away. My family will be secured".

Many participants in the focus groups said that, because the signature of the woman is required after the JLO, getting JLO meant notable changes in how important decisions regarding land were taken, such as transferring, selling, leasing, and mortgaging land. However, in the daily management of land and in crop production, it depends on individual behavior more than on the legal changes regarding land.

JLO status reduces the possibility of becoming landless. One of the respondents who obtained JLO almost a decade back shared that,

> My husband wanted to sell one katha[1] land for NRs 0.2 million few years ago, but I stopped him from selling that piece of land. Due to my JLO status, I could stop him in time. Today the price of that land has increased to NRs 4 million. And we have built one cemented, four room house and are living with comfort.

Reduced Gender-Based Violence

Some other women shared that incidences of gender-based violence had dropped because of their JLO status. Many women spoke about changes at the household level relationship, such as a decline in domestic conflicts. One of the respondents shared that, "I feel secure. The reduced conflicts at home and violence are because of JLO. My husband and sons take care of me more than before; we are all happy with JLO."

Enhancing Women's Access to Financial Resources

JLO provides courage to women to obtain legal access to land and to use the land as a mortgage for bank loans to pursue economic enterprises. A woman JLO owner recently mortgaged her land and took a loan from a bank for five years. With the loan she has started her business of tailoring and clothing store, increasing her income. She said, "With the landownership I could use access bank loan and now I'm the owner of land as well as a business."

Enhancing Women's Access to Health

Women have demonstrated increased access to health services. One of the Brahmin women from Rasuwa district shared that, with JLO, she could make decisions on the food choices and cultivation practices, including the decisions on the use of fertilizer. This has improved her and her family's well-being and good health.

Discussion and Conclusion

Analyzing the factors affecting the distribution of land ownership, this chapter highlights how JLO is breaking the glass ceiling of customary laws on land ownership. JLO has proven to be a gender transformative mechanism helping women to claim land and assets on an equal footing with that of men. This enables them greater participation -- to share their voices and choices on diversified land use to enhance family cohesion, well-being, and prosperity.

JLO not only deals with the physical and legal access to land but also helps to achieve wider recognition of women as landowners through redistribution of land ownership and increase women's participation in land-related decisions.

Our findings substantiate the concepts and ideas of the Land Rights Forum, which recommends that expediting joint-couple certificates will give due recognition to women, create harmony in the family, and decrease violence against women. It also says that with the status of women improved there will be fewer landless squatters and less fragmentation of land withholdings. By having equal ownership, women are motivated to maximize land use participation, enhancing livelihood and well-being. This can reduce the trend of land fragmentation. JLO can provide control over land, expand women's agency, and increase self-esteem (Klugman et al., 2014).

The JLO provides couples with legal consent over land decisions, but the persistence of patriarchal social norms and culturally biased gender ideologies tend to undermine the effective implementation of land-related laws and policies (Bayisenge, 2018).

This analysis of JLO has contributed to applying an environmental lens to understanding gender-based agrarian disparities in Nepal. In particular, it has shown how JLO leads to distributive justice in increasing women's access to land and related productive resources. JLO promotes procedural justice by re-positioning women in empowered roles in decision-making on the inheritance, acquisition, use, and sale of land. Finally, JLO is part of a movement towards recognizing and addressing the specific gender (as well as class and ethnic) factors based on hetero-patriarchal cultural legacies that lead to environmental injustices.

Recommendations

More research is needed to explore the policy impacts of JLO, and to establish evidence from a wider base of communities. We also need a better understanding of how women intersected by age, ethnicity, class have access to land ownership and examining the pros and cons of joint land ownership. Additionally, there is scope for further recognizing women's lineage in the landownership and redistribution of land ownership between men and women for achieving justice and meeting sustainable development goals. Likewise, JLO research so far has been targeted to smallholders, mostly ignoring wealthy land owners.

To spread the benefits of JLO for gender equality more broadly, several steps need to be taken:

- Civil society needs to strengthen and sustain JLO campaigns through mass awareness of the long-term benefits of gender equality.
- The national and provincial governments must pass laws such that the JLO provision is compulsory.
- The government must integrate JLO awareness and capacity building in all sectoral plans and programs for long-term implementation.

Note

1 One kattha [Nepal] is equal to *338 square meter*.

References

Agrawal, B. (1994a). *A Field of One's Own: Gender and Land Rights in South Asia*. Cambridge University Press.

Agrawal, B. (1994b). Land Rights for Women: Making the Case. In *A Field of One's Own: Gender and Land Rights in South Asia* (pp. 1–50). Oxford University Press.

Akter, S., Rutsaert, P., Luis, J., Me, N., Su, S., Raharjo, B., & Pustika, A. (2017). Women ' s Empowerment and Gender Equity in Agriculture : A Different Perspective from Southeast Asia. *Food Policy*, *69*, 270–279. https://doi.org/10.1016/j.foodpol.2017.05.003

Bayisenge, J. (2018). From Male to Joint Land Ownership: Women's Experiences of the Land Tenure Reform Programme in Rwanda. *Journal of Agrarian Change*, *18*(3), 588–605. https://doi.org/10.1111/joac.12257

Chaudhary, S., McGregor, A., Houston, D., & Chettri, N. (2018). Environmental Justice and Ecosystem Services: A Disaggregated Analysis of Community Access to Forest Benefits in Nepal. *Ecosystem Services*, *29*(October 2017), 316–332. https://doi.org/10.1016/j.ecoser.2018.01.009

Fraser, N. (1997). From Redistribution to Recognition? Dilemmas of Justice in a "Post-socialist" Age. In *Justice Interruptus* (pp. 21–50). https://doi.org/10.4324/9781315822174-8

García-Morán, A., & Yates, J. S. (2022). In between Rights and Power: Women's Land Rights and the Gendered Politics of Land Ownership, Use, and Management in Mexican Ejidos. *World Development*, *152*. https://doi.org/10.1016/j.worlddev.2021.105804

GoN. (2015). *Nepal's Constitution of 2015*. Government of Nepal.

GoN. (2021). *Natioanl Population and Housing Census 2021, National Report*. Government of Nepal.

IOM. (2016). Barriers to Women's Land and Property Access and Ownership in Nepal. *International Organisation for Migration*, 75. http://publications.iom.int/books/barriers-womens-land-and-property-access-and-ownership-nepal

Keera, A. (2007). Do Women's Land Rights Promote Empowerment and Child Health in Nepal? *World Development*, *1*(11), 1975–1988. https://doi.org/10.1016/j.worlddev.2006.12.005

Klugman, J., Hanmer, L., Twigg, S., Hasan, T., Mccleary-Sills, J., & Santamaria, J. (2014). In *Voice and Agency: Empowering Women and Girls for Shared Prosperity*. Washington: World Bank.

Koirala, S. (2022). Women's Land Ownership and Gender Equality in Nepal. *Journal of Applied Social Science*, *16*(2), 533–547. https://doi.org/10.1177/19367244221077624

Rawls, J. (1971). *A Theory of Justice*. Cambridge (Mass.).

Schlosberg, D. (2004). Reconceiving Environmental Justice: Global Movements and Political Theories. *Environmental Politics*, *13*(3), 517–540. https://doi.org/10.1080/0964401042000229025

Tamang, S., Paudel, K. P., & Shrestha, K. K. (2014). Feminization of Agriculture and its Implications for Food Security in Rural Nepal. *Journal of Forest And Livelihood*, *12*(1), 20–32.

UN Women. (2020). *Realizing Women's Rights to Land and Other Productive Resources* (Second Edition), New York: United Nations.

Villa, M. (2017). *Gender Inequality Women Own Less than 20% of the World's Land. It's Time to Give them Equal Property Rights*. World Economic Forum. https://www.weforum.org/agenda/2017/01/women-own-less-than-20-of-the-worlds-land-its-time-to-give-them-equal-property-rights/

WOM. (2018). *Gender Equality in Land Rights, Ownership Vital to Realizing 2030 Agenda, Women's Commission Hears amid Calls for Data Collection on Tenure Security*. Meetings Coverage and Press Releases Sixty-Second Session, 10th & 11th Meetings on the Commision of the Status of Women. https://press.un.org/en/2018/wom2143.doc.htm

Young, I. M. (1990). *Justice and the Politics of Difference*. Princeton University Press.

6 Environmental Justice and Unfree Agricultural Laborers in the Eastern Tarai of Nepal

Suresh Kumar Dhakal

Introduction

A government report[1] estimates that 69,000 Nepali families living under the status of Harawa-Charawa (HC), or unfree/ bonded Dalit labor in Nepal, are grossly denied environmental justice (EJ). However, the literature on EJ in Nepal has yet to critically assess the disproportionate impact of the hazards on the life and livelihood of HC and similar enslaved labor populations. Previous studies on HC and other forms of modern slavery in Nepal have focused primarily on economic dimensions (Burns et al., 2016; KC et al., 2013; Giri, 2010; Upadhya, 2008; Sharma and Shanna, 2002; Robertson & Mishra, 1997; NNDSWO & LWf, n.d.). Comparatively, the environmental inequalities faced by the unfree/bonded agricultural laborer are less documented.[2] To address these shortcomings, this chapter uses ethnographic and survey data to examine the inequitable impacts of environmental hazards on HC, specifically in the Madhesh province in Southeastern Nepal which has experienced the highest cases of climate-driven losses and damages in recent times (Shrestha et al., 2020).

If EJ is "a solution to environmental inequality and injustice" (Pellow, 2000), assessing the impact of environmental change on HC can expand the scope of critical EJ studies. Since "environmental inequality focuses on broader dimension of the intersection between environmental quality and social hierarchy" (Pellow, 2000, 582), this chapter suggests a thorough examination of environmental inequality as a form of structural inequality and environmental burden. This chapter hopes to contribute to Pellow's (2017) first pillar of critical environmental justice and intersectionality by documenting how caste, ethnicity, and class place enslaved labor at the bottom of a complex system of hierarchies associated with Nepal's agricultural system. This, in turn, places these populations at extreme risk for climate-induced hazards with little recourse to legal relief. Hence, the chapter joins recent work to integrate slavery and the socio-ecological landscape of the plantation as a key context for the historical and present state of environmental injustice (Brown, 2021; McKittrick, 2011, 2013).

By examining the interlinkages between landlessness, socio-economic inequalities, unequal impacts of climatic hazards, and the right to protection against injustice, this chapter concludes that historically constructed social and economic discrimination and vulnerability are the bedrock of environmental injustice of

DOI: 10.4324/9781003371175-8
This chapter has been made available under a CC-BY-NC-ND 4.0 license.

unfree agricultural labor, and this understanding could be better enhanced through the concept of environmentalism of the poor.

Harawa-Charawa, the Unfree Agricultural Laborer: Data and Method

Nepal's HC system is often associated with modern slavery. The work contracts governing HC agricultural laborers are often referred to as unfree (Dhakal, 2007; Brass, 1999), bonded (Damir-Geilsdorf et al., 2016; Fudge & Strauss, 2014), indentured servitude (Suranyi, 2021), new slavery (Bales, 1999), or forced labor (ILO, 2023). In this chapter, considering the context we propose a definition of HC, which we created after extensive conversations with the various stakeholders, which is as follows,

> An individual coerced into working in a landlord's field as a plowman or any other assigned agricultural work for the interest of the loan received, or for grains or the small piece of land cultivated or for shelter in a bonded situation is a Harawa. Likewise, a person who does not have the freedom to leave work or has no choice but to work the field and does not have a fixed working hour, receives insufficient or no wage (i.e., either in cash or in-kind) for his labor is also called a Harawa.

The HC system emerged alongside Nepal's feudal land systems starting in the 1700s, which granted land to its functionaries and service providers, members of the royal family, warrior family, priests, and close kins and officials (Regmi, 1978). Through such a state-facilitated process, a few privileged families accumulated large areas of land, while others were forced to provide free service, known as *Jhara,* to those landlords. Over the next few centuries, this unfree labor system was mapped onto Nepal's caste system, which positioned members of the most oppressed occupational caste (Dalits) at the bottom of this hierarchy (Dhakal, 2007).

This in turn has safeguarded access and power to the 'higher caste'. For example, access to land has been guaranteed only to the 'high caste' people; the majority of the Dalits have remained landless and must depend on landlords for access to agricultural and residential land as well as cash for basic needs (Gautam, 2017; Iverson, 2013; Dhakal, 2011).

HC could not detach themselves from their landlords as they feared losing their shelters and the land they cultivated. This land arrangement is called *Laguwai*.[3] Others fear losing due wages or their share of the crops and physical abuse. Traditionally, HCs would enter into a contract through a certain ritualized process. Hence, HC is wrapped up in social and cultural fabrics and has economic dimensions (Dhakal, 2007).

Methods

This chapter draws from the survey conducted in Charnath Ganeshman Municipality, Shahidnagar Municipality, and Dhanauji Rural Municipality in the Dhanusha district of Madhesh Province in 2022 (Dhakal et al., 2020). Here, 3,636 (16.7%) of the total 19,425 families were HC; 1,925 (53%) of whom were in debt bondage and were extremely vulnerable. One-third (32%) of Terai Dalits working as HCs had inherited the position from their parents and were forced to continue as one.

Combining and comparing the data from the surveys (Dhakal et al., 2020; Dhakal, 2007) and ethnographic accounts collected in several follow-up field visits between 2018 and 2023, this chapter shows that HC's weak socioeconomic position makes them susceptible to environmental hazards.

Caste, Class, and Land

Historically, the caste-class dynamics have complicated the issue of bonded labor in South Asia. The question of the environmental injustice of HC should be understood in the context of historical development, as well. "Environmental injustice is not just a single harmful event/action/result, but rather a complicated history of political, social, and economic interactions leading up to, and continuing beyond, the contested instance of perceived injustice" (Sze & London, 2008: 1333). As such, the historical caste-class interface has also shaped the present practice of HC.

Due to socio-economic discrimination, the HCs, who are mostly Dalits and landless, face a disproportionate impact of climate change and have the least adaptive capability to protect themselves from climate-related floods and other hazards (Murphy-Greene, 2022). A majority (88%) of HC surveyed live in temporary shelters made of wattle and daub, and only 29% of them have ownership of the land where they have built their house. The rest have been built on *Gaaun Block*[4] (27%), on unregistered land or public land (26%), and *guthi* institutional land (5%). Usually, HC settlements are clustered in congested areas, and spaces inside and outside of houses are crowded and are of low-quality construction. HCs have limited food security, with more than one-third (39%) of them being landless. Another 27% can survive up to three months from their agricultural produce, 26% for up to six months, and 7% for up to nine months. Only 1% said they produce enough to last the whole year.

The DesInventar database shows that from 1971 to 2016, a total of 4,160 flood events hit Nepal, of which 26.5% were in Madhesh Province. It also shows that the damages these floods caused were highest in Madhesh. In recent years, too, Madhesh Province has experienced a record number of floods and resulting losses and damages leading to a huge number of environmental victims who are mostly the poor people living in the floodplain. For example, Madhesh accounted for 45% (4,160) of all damaged houses in Nepal in 2020. Likewise, it constituted 70% of Nepal's total population affected, and economic loss was also the highest at 45.14% (Shrestha et al., 2020).

We documented that, in the last few rainy seasons in Madhesh Province, settlements had been flooded, shelters collapsed, and stored food was destroyed. In some cases, HC families had to take refuge in local schools. During such events, HC mobility was restricted, and they often lost work and wages. Children could not go to schools as culverts were flooded and trails were washed away. Climate change also creates temperature hazards as many infants and elderly people die of extreme cold in the winter and farmers often get sick or die from heat stroke.

HCs are not only vulnerable to storms and other hazards but often receive inadequate relief. A Dalit woman told us, "Even if Dalits are affected most by the heavy rain and flooding, relief support is always delayed and we always get it last. We are weak; therefore we are discriminated against."[5] A rights activist working with HCs explained that because of a "lack of access to relevant information, the response is delayed in some cases." The Vice-Chairperson of Dhanuji Rural Municipality further clarified that

> the *Palika* itself does not discriminate, but those who are affected cannot approach the office by themselves on time. Whoever is responsible for distributing relief and rescue activities may have been late to get there also because of the limited human and other resources.[6]

As climatic events such as flooding, drought, and other extreme weather become more frequent and severe (Awasthi & Owen, 2020), HCs and other marginalized groups will experience worsening living and working conditions. During our fieldwork in different HC settlements in Dhanusha, people reported that the incidences of natural disasters and the risks to HC have increased in recent years.

Nepal's Monsoon Prepared and Response Plan 2079 BS (2023) highlights the severe risk of floods and the lack of preparedness in the HC settlements. While the plan sets forth several strategies to address these risks, there is little progress on the ground in HC settlements. One local woman said, "Whatever happens, happens without prior information, nobody visits our Musahari Tole (a Dalit group) and warns us about the risk" (Figure 6.1).

Figure 6.1 A mother and daughter in front of their flood-damaged home, Dhanusha District.
Photo credit: Author.

Climate change also is reducing access to forests and other natural resources. This makes HCs more vulnerable, as 68% of HC households still use firewood, and 30% of them use dried cow dung as the main source of cooking fuel.

The State's policies on land and resource access exacerbate climate injustice. Had there been a pro-poor land reform that protected the tenure security of the poor and marginalized families, such injustices could have been reduced. To be sure, the 7th and 8th Amendments of the Land Act of 1964 have provisions on land for Dalits and landless people, and the 18th Amendment of the Land Act outlines the amount of land to be entitled to each of those families. If these amendments were implemented, the dignified rehabilitation of the Dalits could have been accomplished, and much of the loss and damages due to environmental adversities today could have been prevented.

The government's unwillingness to implement the available labor and land policies has also obstructed the process of ensuring EJ for HCs, and other landless marginal groups. The Bonded Labor Prohibition Act (2002), Constitution of Nepal (2015), National Land Policy (2019), Land Use Policy (2015), and Land Use Act (2019) including several amendments to the Land Act, and Labor Act (2017) and amendments related to the minimum wage are some of the bases for the legal actions. In addition, several Articles of the Constitution 2015 pertain to social and economic rights[7]; and the Directive Principles, Policies, and Responsibilities (Part 4) regarding social justice and inclusion, require the federal government to ensure the rehabilitation of *Kamaiya* (bonded laborers), *Kamlari, Haruwa, Charuwa, Haliya*, the landless and the squatters by identifying them, and making arrangements of housing or providing a small plot of land or house, employment or arable land for their livelihoods.

Delayed or denial of implementation of the available policies is a lost opportunity in addressing climate-induced environmental injustices affecting HCs and similar groups.

Land, Labor, and Environmental Justice

The complex political-economic relations that have marginalized HC can be better understood and addressed through the framework of the 'environmentalism of the poor' (Folchi, 2019; Ghimire, 2003; Guha & Martínez-Alier, 1997; Nixon, 2011). HCs are tied to their landlords for survival and have historically been exposed to environmental and social risks. Environmentalism of the poor also recognizes the 'locally rooted environmental ideology' as useful insights regarding the ecological and other wider concerns (Guha, 1991). Therefore, EJ cannot be studied in isolation or evenly across social groups, but rather requires a contextual intersectional approach linking caste and class.

An intersectional approach linking historically constructed land-labor relations better explains the EJ of the rural poor such as the HC groups. And, it aims at 'enhancing and improving the livelihood of the poor, less privileged, minority, and affected people sustainably since they have depended directly on natural resources for substance' (Ghimire, 2003). Hence, this conforms to 'materialistic and based on interest' environmentalism, as in other countries of the global South (Guha & Martinez-Alier, 1997).

The policies and mechanisms to ensure meaningful participation of the most affected population, HC in this context, in the decision-making processes are grossly lacking. During an interaction program at Dhanusha, representatives of HC expressed their discontent at never being consulted nor involved in the decision-making process, even at the ward level. They noted that there was no practice or mechanism for informing HC and marginalized families even about mandatory meetings, like the ones conducted by *tole bikas samiti*, a standard procedure of local governments to collect ideas for development plans. This orients us to think that the movements of HCs for secured access to land and other natural resources, access to information, meaningful participation in the decision-making processes, and organized resistance against environmental injustices must be integrated into EJ studies.

Since environmental risks, damages, and losses are unequally distributed by social and economic class, the 'linkages between slavery and environmental changes' should be considered as a distinct branch of research in EJ studies (Brown et al., 2021), where the notion of justice has to be grounded in knowledge and experience and to develop out context (Sultana, 2015). Therefore, environmental injustice should be addressed based on the knowledge and experience of the people living on the social, economic, and ecological margins, like HCs.

Altogether, social and economic discrimination based on caste and class faced by HC has placed them at risk of climate-induced stressors and disasters. HCs are not politically organized, particularly around the issues of exploitation and injustice, and are not economically stable enough to effectively demand their legal rights and broader social and environmental justice.

Conclusion

The suffering of HCs from unequal impacts of extreme weather events linked to the climate crisis is a byproduct of historical injustices. Those suffering economic and social inequalities are readily exposed to uneven environmental vulnerabilities. But they lack financial and social capabilities, therefore HCs along with other landless and marginal groups are far behind in collective actions in resisting environmental injustice and other forms of discrimination and exploitation.

The HC status creates distinct challenges relative to environmental hazards and in particular climate change, as its impacts are disproportionate leading to unequal damage and loss. Such disproportionate impact owing to their class and caste position makes it an even more complicated issue that can't merely be addressed with a general framework; rather we need an effective mechanism to ensure EJ for the HCs. Any policy to address the injustice induced by the disproportionate impacts, which is unequal damage and loss, must link caste, class, labor, agriculture, and environmental aspects.

Finally, unfree/bonded agricultural laborers are a distinct and important category of marginalized populations that EJ studies need to incorporate, which can appropriately be brought in through the concept of environmentalism of the poor.

Recommendations

Implementation of existing policies is the first step for the resettlement and rehabilitation of HCs with tenure security to reduce further disproportionate damage and loss caused by the climate crisis. By assessing such disproportionate negative impacts in the forms of damage and loss, the government at different tiers should determine and allocate the required resources and take actions to reduce the adverse impacts of climate change and ensure justice for HC and other marginalized groups. Policies formulated with meaningful participation of the affected populations must link caste, land, labor, and environmental aspects to address the issue holistically.

Notes

1 *Mukta Kamaiya, Kamlhari, Haliya ra Harawa-Charawako Bastusthiti Adhyan Report 2078* (A Study on Situation of Liberated Kamaiya, Kamlhari, Haliya and Harawa-Charawas. The study committee was formed by the Nepal Government's Ministry of Land Management, Cooperatives, and Poverty Alleviation.

2 HC practice is considered a form of modern slavery. For example, according to the definition by Anti-slavery International, the HC system of Nepal possesses a feature of forced labor and debt-bondage/bonded labor. https://www.antislavery.org/slavery-today/modern-slavery/.

3 HC families are provided with a piece of land, often degraded land and less productive, which they can cultivate as they serve as HC to the respective landlord.

4 A parcel of land registered as collective ownership by several individual families, no individually delineated land as an individual plot, and hence no ownership certificate is issued individually. This is particularly prevalent in Madhesh Province.

5 Field interviews on 26 December 2022 in Musahari Tole, Dhanauji, Dhanusha.

6 Interviews from an interaction program in Dhanusha on Friday, 29th December 2022.

7 Right to live with dignity (Article 16): Each person shall have the right to live with dignity.
Right to Freedom (Article 17): Every citizen shall have the freedom to engage in any occupation or be engaged in any employment, trade, or business in any part of Nepal.
Right to Equality (Article 18): All citizens shall be equal before the law. There shall be no discrimination based on caste, religion, race, sex, tribe, physical conditions, or any other such grounds.
Right against Exploitation (Article 29): Every person shall have the right against exploitation. No person shall be subjected to any kind of exploitation based on religion, custom practices, or others. No person shall be subjected to human trafficking or bonded labor, and such an act shall be punishable by law. No person shall be subjected to forced labor.
Right regarding Labor (Article 34): Every laborer shall have the right to proper work practices. Every laborer shall have the right to appropriate remuneration; facilities and contribution-based social security.

References

Awasthi, R. and Owen, J. (2020 Dec.). "Observed Climate Extreme in *Nepal*", *The Geographic Base* 7: 1–14.

Bales, K. (1999). *Disposable People: New Slavery in the Global Economy.* Berkeley: The University of California Press.

Brass, T. (1999). *Towards a Comparative Political Economy of Unfree Labour: Case Studies and Debate.* London: Frank Cass.

Brown, D., Boyd, D. S., Brickell, K., Ives, C. D., Natarajan, N., & Parsons, L. (2021). Modern Slavery, Environmental Degradation, and Climate Change: Fisheries, Fields, Forests, and Factories. *Nature and Space* 4(2): 191–207.

Burns, D., Dharma, B. P., & Oosterhoff, P. (2016). *Patterns and Dynamics of Bonded Labor, Child Labor, and Child Marriage in the Nepali Eastern Terai: Findings from Life Story Analysis.* Brighton, U.K.: Institute of Development Studies.

Damir-Geilsdorf, S., Lindner, U., Müller, G., Tappe, O., & Zeuske, M. (eds.). (2016). *Bonded Labour: Global and Contemporary Perspective (18th – 21st century).* Bielefeld: transcript Veriag.

Dhakal, S. (2007). Haruwas, the Unfree Agricultural Labourer: A Case Study from Eastern Tarai. *Contributions to Nepalese Studies* 34(2): 277–301.

Dhakal, S. (2011). *Land Tenure and Agrarian Reform in Nepal.* Kathmandu: Community Self-reliance Centre.

Dhakal, S. Karki, K. & Shrestha, S. (2020). Harawa-Charawa, the Bonded Laborers in Agriculture. A Study of Three Municipalities of Dhanusha District in Eastern Terai. Kathmandu: Community Self-Reliance Centre.

Folchi, M. (2019). Environmentalism of the Poor: Environmental Conflicts and Environmental Justice. In Delgado, L., & Marín, V. (eds.), *Social-ecological Systems of Latin America: Complexities and Challenges.* Cham: Springer. https://doi.org/10.1007/978-3-030-28452-7_6

Fudge, J., & Strauss, K. (eds.). (2014). *Temporary Work, Agencies and Unfree Labour: Insecurity in the New World of Work.* New York: Routledge.

Gautam, T. R. (2017). Ethnicity and Land Ownership in Nepal. *Contemporary Social Sciences* 26(3): 123–136.

Ghimire, S. (2003). "Concept of Environmental Justice in Nepal: Environmentalism of the Poor for Sustainable Livelihood." *Himalayan Journal of Sciences, 1*(1), 47–50.

Giri, B. R. (2010). The Bonded Labor System in Nepal: Exploring Halia and Kamaiya Children's Lifeworlds. *Himalaya* 29(1): 29–41.

Guha, R. (1991). *The Unquiet Woods: Ecological Change and Peasants Resistance in the Himalaya.* New Delhi: Oxford University Press.

Guha, R., & Martínez-Alier, J. M. (1997). *Varieties of Environmentalism Essays North and South.* London: Routledge.

Iverson, T. O. (2013). *Poverty Dynamic among the Dalits of Tarai.* Master thesis to complete the degree of Master in Economics, University of Bergen.

KC, B. K., Subedi, G., & Suwal, B. R. (2013). *Forced Labour of Adults and Children in the Agriculture Sector of Nepal: Focusing on Harwa-Charuwa in Eastern Tarai and Haliya in Far-Western Hills.* Kathmandu: International Labor Organization, ILO.

McKittrick, K. (2011). On Plantations, Prisons, and a Black Sense of Place. *Social and Cultural Geography* 12(8): 947–963.

McKittrick, K. (2013). Plantation Futures. *Small Axe: A Caribbean Journal of Criticism* 17(3): 1–15.

Murphy-Greene, C. (2022). An Overview of Environmental Justice. In Murphy-Greene, C. (ed.), *Environmental Justice and Resiliency in an Age of Uncertainty.* New York: Routledge (pp. 4–13).

Nixon, R. (2011). *Slow Violence and the Environmentalism of the Poor.* Cambridge: Harvard University Press.

NNDSWO & LWF (n.d.). Haliya Survey in Far Western Development Region. Nepal: NNDSWO/LWF.

Pellow, D. (2017). *What is Critical Environmental Justice?* New York: John Wiley & Sons.

Pellow, D. N. (2000). Environmental Inequality Formation: Toward a Theory of Environmental Injustice. *American Behavioral Scientist* 43: 581–601.

Regmi, M. C. (1978). *Land Tenure and Taxation in Nepal.* 2nd Edition. Kathmandu: Ratna Pustak Bhandar: Himalaya.

Robertson, A., & Mishra, S. (1997). *Forced to Plough: Bonded Labor in Nepali Agricultural Economy.* London: Anti-Slavery International.

Sharma, S., & Sharma, R. K. (2002). *Long-term Farm Labour Systems in Nepal.* Kathmandu: National Labour Academy-Nepal. A report submitted to International Labour Organization.

Shrestha, B. R., Rai, R. K., & Marasini, S. (2020). Review of Flood Hazards Studies in Nepal. *The Geographic Base* 7: 24–32.

Sultana, F. (2015). Justice. In Agnew, J., Mamadough, V., Secor, A. J., & Sharp, J. (eds.), *The Wily Blackwell Companion to Political Geography* (pp. 127–139). Chichester: Wiley Blackwell.

Suranyi, A. (2021). *Indenture Servitude: Unfree Labour and Citizenship in the British Colonies.* Montreal & Kingston: McGill–Queen's University Press.

Sze, J., & London, J. K. (2008). Environmental Justice at the Crossroads. *Sociology Compass* 2(4): 1331–1354. Doi: 10.1111/j.1751–9020.2008.00131.x.

Upadhyaya, K. P. (2008). *Poverty, Discrimination, and Slavery: The Reality of Bonded Labor in India, Nepal, and Pakistan.* London: Anti-Slavery International.

7 Connecting Dalit Land Rights and Climate Justice

Madan Pariyar and Arjun Kumar B.K.

Introduction

Dalits in Nepal, as elsewhere, have historically suffered from the atrocities of untouchability and caste-based discrimination. Comprising approximately 14% of Nepal's total population of nearly 30 million, they are socially, economically, politically, and culturally excluded. The 1854 Country Code (*Muluki Ain*), which established the caste-based discrimination and untouchability system in Nepal, was in force for a century. Similar to other Hindu-dominated countries in South Asia, Nepal's Code reconstructed social structure into a caste hierarchy, placing Hindu Brahmins and Chhetris at the top and Dalits – the groups belonging to the *pani nachalne chhoi chhito halnu parne* (castes from whom water is not accepted and whose touch requires purification by sprinkling water) category – at the bottom.

Caste discrimination is an important, yet understudied, dimension of environmental and climate justice in South Asia and the South Asian diaspora. Dalits confront inequitable impacts of environmental and climate change-related stressors and marginalization from decision-making to mitigate and adapt to these impacts (Ranganathan, 2022; Sharma, 2022). This chapter will contribute to this nascent literature through an exploration of the historical and contemporary implications of caste-based discrimination on the health and well-being of Dalits in Nepal in a time of climate crisis.

In Nepal, the practice of untouchability was implemented through provisions for differential caste-based social restrictions and punishments. Over the last 60 years several legal reforms and policy initiatives have been undertaken, including the New Country Code 1963 (*Naya Muluki Ain*), the 1990 Constitution, the declaration of Nepal as a country free from caste-based discrimination and untouchability on June 4, 2006, the Interim Constitution 2007, the Untouchability and Discrimination Act 2011, and the Constitution of Nepal 2015. All these legal instruments broadly aim at abolishing caste-based discrimination and untouchability, and at guaranteeing that all citizens in Nepal are equal irrespective of caste, creed, and gender.

Despite these legal provisions, Dalits today are still facing severe discrimination and violence, and are deprived of access to resources and infrastructure necessary for health and well-being. This includes marriage segregation, inter-caste marriage violence, denial of access to temples, religious and cultural functions, denial of

DOI: 10.4324/9781003371175-9
This chapter has been made available under a CC-BY-NC-ND 4.0 license.

access to public utilities and educational facilities, denial of clean water and other natural resources, blocking access to housing, and occupational segregation. Many of these factors are exacerbated by the worsening impacts of the climate crisis.

The country's prevailing laws and practices have deprived Dalits of land entitlements and possessions of other assets and properties. Most Dalits, consequently, are landless. Various studies reveal that landlessness among Dalits is extreme, approximately 36.7% among the hill Dalits and 41.4% among Madheshi Dalits (lowland Dalits). Those that do hold land tend to have very small landholdings barely sufficient for subsistence, never mind income generation.

The chapter will address questions of landlessness and marginalization in the context of climate justice through a case study that will document and analyze the experiences of landless Dalits. The chapter draws on pertinent secondary literature along with a review of relevant legal documents, interactions with knowledgeable persons/stakeholders, and on-site visits to Dalit settlements to acquire firsthand information on issues related to Dalit land rights and land entitlements.

Background: Inequality in Access to Land

Dalits' landlessness is the result of historic suppression and exclusion imposed by the state on the grounds of Hindu Varna Vyavastha and caste systems. The millennia-old Hindu Varna Vyavastha system has restricted Dalit land ownership and led to landlessness and insecure tenure (Dulal and BK, 2019). Similarly, caste-based discrimination has led to these communities being treated as untouchables in the Nepali society for centuries. Guided by the myths that their touch will contaminate privileged caste human bodies, temples, water taps, and other services/facilities, Dalits have been marginalized from society, and have restricted from securing land and other properties.

The land tenure system in Nepal has a historical basis in feudal landlordism. Until 1951 the autocratic Ranas granted most of Nepal's fertile and prime land to courtiers and various elites. Because various land reform efforts were ineffective, this unequal system of land tenure continued under the return of the Shah monarchy that lasted until the monarchy was overthrown in 2008. Peasants, the real tillers of land, have been left behind with limited access and poorly recognized tenure rights. Landlessness has made Dalits economically vulnerable and dependent upon landlords. Such dependency compels them to cultivate land under exploitative tenure arrangements. The root causes of share cropping, haliya, balighare/khalo, charuwa haruwa systems are associated with the issue of landlessness and small holding (Nepali, 2008; Giri, 2009; Dhakal, 2011). Today, both poor landless urban dwellers and peasants in informal settlements, who are mostly Dalits, face constant fear of eviction.

Land continues to be the most significant asset in Nepal's rural-agrarian economy. Nearly two-thirds of the country's population depend on agriculture and allied pursuits for their livelihood. However, most of the land is concentrated in the hands of the few. The richest 7% of households own about 31% of agricultural land. The size of holdings does not support subsistence or effective engagement in the market economy. In 2002, some 47% of the Nepali farmers owned less than 0.5 hectare of land. This marginal ownership status increased to 54% by 2012 (CBS, 2013).

Nepal has 569,400 farming families classified as 'landless.' This means that they either lack formal documentation for the land that they have been living and farming on for generations, or they have no land at all and are either tenant farmers or are farming on government land without express permission (IIED, 2020). The latest estimate shows that about 25% of the arable land with an estimated 10 million properties are lacking legal tenure status. Around 1.34 million households are living under informality and are under constant fear of eviction (Pandey et al., 2021). Nearly 5–7 million urban poor are landless squatters and have even lost their daily livelihoods in 2020 due to the Covid-19 pandemic (NHDR, 2020).

While they comprise about 14% of the Nepal's population, Dalits own only 1% of the arable land. Landlessness among Dalits is extreme: 36.7% among the hill *Dalits* and 41.4% among Madheshi Dalits. Those Dalits that do hold land tend to have very small landholdings. For example, 77% of Dalit families in the hills and 95% of Dalit families in the Terai possess less than 0.1 hectare of land. Dalits often do not have legal title to their lands. This weakens their tenure security and makes it difficult to bequeath land to their children. It also creates challenges when applying for essential services such as getting their homes connected to the electrical grid. As a result of their limited landownership, Dalits comprise the poorest of the poor in Nepal, in terms of income, consumption and human development. A total of 42% of Dalits fall below the poverty line compared to the national average of 25.2%.

This uneven allocation, poor access and unsecured tenure of vulnerable groups over land has triggered political unrest, civil society movements, and even armed conflict. The demand for access to land to the landless and tenure security for all was one of the triggers of the armed conflict between the Maoists and the government between 1996 and 2006. After the Comprehensive Peace Agreement (CPA) in 2006, a national consensus emerged to attempt a resolution of the historical injustice in land governance. The new Constitution promulgated in 2015 and the Local Government Operation Act of 2017 have granted land rights. But these provisions are yet to be fully realized. According to the Ministry of Land Reforms and Poverty Alleviation, only 31,057 hectare of land has been distributed to 154,854 landless/squatter families during the last 30 years (Dhaulakoti, 2022). Equitable land distribution, justice and social inclusion still remain a wishful dream for the landless Dalits of Nepal. Apropos this chapter, landlessness makes Dalits even more vulnerable to the problems brought on by climate change that is already affecting Nepali communities, both in the hills and the lowlands.

Case Study: The Differential Impacts of Climate Change and Climate Injustice on Dalits

Climate change is imposing significant impacts on the human populations and ecosystems. Changes in precipitation patterns have led to more frequent and intense rainfall events, causing landslides and flash floods in both the hills and lowland areas of Nepal. Conversely, reductions in snowpack, due to higher winter temperatures, the drying up of springs and shifting timing of the summer monsoon has left many communities without reliable water supply for drinking and agriculture. Rising temperatures have led to the melting of glaciers and increased snowmelt, resulting in the formation of glacial lakes and the potential threat of Glacial Lake Outburst Floods (GLOFs) and loss of perennial water sources. The agricultural sector, which heavily relies on rainfall and snowmelt for irrigation, has been especially destabilized by changing precipitation patterns, resulting in reduced crop yields and food insecurity. The impacts of climate change in Nepal pose significant challenges to the country's development, economy, food security, and overall well-being of its people, especially Dalits and other marginalized populations.

In 2022, the Samata Foundation, a leading NGO that works on Dalit issues in Nepal, conducted a study titled "Causes of Climate Vulnerability, Conflict Dynamics and Existing Local Adaptive Capacities of Dalit Communities in Nepal" (Pariyar, 2022). The study documented the experiences and perspectives of Dalit communities at the grassroots level in Nepal with regard to climate change and climate justice. The study defines climate justice as finding solutions that not only reduce emissions, but that also creates a fairer, more just and more equitable process for inclusion of the poor and the marginalized people who suffer the most from the impacts of the climate crisis.

The research was carried out in Siraha and Saptari districts of Madhesh province (in Nepal's lowlands), and in Rolpa and Rukum East districts in the hilly region of Lumbini province. The study draws on data collected from 133 households belonging to 21 community-based organizations spread over 16 municipalities/rural municipalities of the aforementioned four districts, 21 focus group discussions involving 252 participants from the grassroots level, and 147 key informants representing different stakeholders at local, provincial, and federal levels.

The study uncovered several important findings as follows.

Climate Change Hazards and Risks: A wide range of climate-induced hazards and risks affect the health and well-being of Dalits. Drought has led to a decline in food production, dried up water sources, and caused human health issues. Floods have eroded riverbank settlements, damaged infrastructure and property, decimated ecosystems necessary to support local livelihoods, destroyed livestock and agricultural lands, and led to the loss of human lives. Landslides have similarly damaged settlements, infrastructure, roads, and ecosystems. Harsh environmental conditions have added to pre-existing food insecurity and increased the mortality rate.

Causes of Climate Vulnerability: Because of their systemic landless and near-landless state, Dalits are more vulnerable to natural disasters than any other caste groups. They are highly exposed to natural disasters because of the poor quality of their housing (often with mud walls and thatched roofs) and their settlements in the margins. In the Hills, Dalits have been forced to settle on the fragile slopes susceptible to landslides, and in the Terai, they live on riverbanks which are at a high risk of flooding and erosion. They are also highly sensitive to disasters because of the nature of their low-wage occupations coupled with poor working conditions, often in exposed outdoor settings given the menial jobs that they undertake. Their adaptive capability is very low because they do not possess the economic resources or political capital to absorb and adapt to the shocks and stresses of disasters. Natural disasters have also worsened Dalit landlessness through the losses of their agricultural land, houses, and other properties (Pariyar, 2022).

One respondent named Kamal Nepali from Rukum East – and now living in the market town of Chhipridah – stated that he wants to go back to his ancestral home but is worried about frequent landslides. Back in his village, Kamal had built a two-story house on a steep slope, not so different than one of his relatives. A year later, half of his relatives' house was swept away by a huge landslide. Only his immediate family survived. Kamal's family has now started a sewing and cutting shop in Chhipridah but he wants to take this business to his hometown when he can also continue farming. Instead of paying expensive rent in the city only to live in a small, rented apartment, he and his family want to go home, but climate-induced landslides have displaced them.

Drivers of Climate Vulnerability: The drivers of climate vulnerability of Dalit communities are poverty and poor economic conditions, voicelessness (inability to put forward their grievances/sufferings to authorities/agencies), low access to resources, lack of political influence, caste-based discrimination, poor social network (isolated and ostracized by the society at large), lack of education and knowledge about climate change induced risks and disasters, geographical remoteness (no roads in the hills), and unpreparedness. In a multiple-choice questionnaire that asked respondents to identify the causes of the climate crisis, 14.6% of them pointed to poverty, 11.2% to caste untouchability, 10.5% to voicelessness and 10.5% to lack of resources.

Coping Strategies Adopted by Dalits: The coping strategies adopted by Dalits include seasonal migration to other cities in Nepal; temporary migration to India, the Middle East and other countries for employment; growing drought/flood-tolerant crops; temporary migration to safer places prior to or during disaster; selling of assets, livestock or other properties, such as ornaments; and seeking safe storage of grains during droughts and floods.

The situation is worse among Madheshi Dalits who are doubly marginalized. Bishewor Prasad Rajak, the head of Madheshi Dalit Federation (MDF) in Rajbiraj, Saptari district said, "Dalits are in very vulnerable conditions. Their settlements are in vulnerable areas. Most of them are settled on the riverbanks, and their houses are built from mud (*Fus Ko Ghar*)" (Figure 7.1).

Existing Local Adaptive Capacities: Both in Terai and hills, Dalits stockpile grains and other food items to be used during the fallow winter seasons. In the Terai, the base level of the houses is raised in order to protect them from flooding. Before the rainy season, the houses are repaired with mud and bamboo.

Figure 7.1 Low-quality housing materials lead to housing damage by storms and floods in the Sakhuwanankarkatti Rural Municipality of the Siraha district.

Photo credit: Usha Karn.

In the hills, the roofs are repaired with thatch from community forests nearby or corrugated sheets purchased from the market. The walls are reworked with mud and stones. Bamboo, gabion, and sand-filled sacks are placed on riverbanks to create temporary embankments as protection against floods. The sheds of animals and birds are shifted to safer places. The community specifies a safe area (mostly schools) for shelter during floods and landslides.

Dalits are also making various efforts to fight these disasters and their aftermath. Dalits of Kalyanpur area in Siraha district are being supported by the National Chure Forest Protection Program to build embankments on the riverbanks. As such, the village is now protected from being flooded. Even so, every now and then their arable lands are covered in thick layers of sand, and unable to grow crops, they face severe food crisis. The Maa Bhagwati Women Farmers Group, which has 25 members, got together and decided to remove the sand accumulated in the fields and use organic fertilizers. Subsequently, the fertility of the soil increased and they managed to grow crops on the same land.

During this study, we found many efforts undertaken by the Dalit community to fight climate disasters at the local level. Some of these include discussing issues related to water and climate change in collaboration with various organizations, mobilizing groups to solve problems, and putting pressure on the local government.

According to Kushmi Devi Ram, the president of the Barhaman Thakur Baba Women's Farmers Group of Siraha, Lahan municipality, the climate impact mitigation program with the support of Samata Foundation and Dalit Jankalyan Yuva Kalb Lahan has taught them many mitigation measures such as sowing ash in the field during the cold wave and planting water-tolerant rice seeds during the monsoon. The community is now cultivating land rented by contract. The use of home compost and cow dung in an otherwise fallow land has increased productivity and yield.

Likewise, the Secretary of Climate Change Mitigation Group of Lahan, Ram Saday said that his group has submitted a group-level plan to mitigate climate change in the municipality following the program. The group was successful in persuading the municipality to allocate a budget of Nrs 100,000 to conduct skill-based training for improving the livelihood of the community. They also got Nrs 20 million from Lahan Municipality to build a two-story building for 38 disadvantaged households.

Differential Impacts of Climate Change: Dalits are not only more vulnerable prior to climate-induced natural disasters, their marginalization means they face discrimination during the rescue, relief, and rehabilitation during and after climate-induced disasters. For example, the distribution of relief packages isn't equitable. During the field survey, 66.7% of the respondents stated they did not get necessary rescue, relief, and rehabilitation assistance during the time of disasters. Likewise, 78.6% of respondents said that they did not get enough relief and support because of their Dalit status. Out of the four districts studied, the number of such under-served people was highest in Siraha district.

One respondent in Siraha district said that Dalits are ignored by the local leaders, and that they rarely visit hard-hit communities or consider their needs and concerns. According to a non-DHRD (Dalit Human Rights Defender) in Siraha district, most of the Dalits, who work under landlords, are suffering from climate-induced food crises. Women face special difficulties as they can't find dry spot to cook, and face difficulties in accessing safe drinking water.

Because of Dalits' low level of material and political capital resources the climate change-related losses and damages are much higher than that of many non-Dalits who may have access to other properties to buffer the climate shocks.

Problems of Climate Recovery Programs: The study points to multiple cases of climate injustice resulting from inequitable distribution of relief materials during and after disasters. It also shows that governments do not have disaster preparedness or recovery plans focused on Dalit communities. Umesh Bisunke, who is the chairperson of Dalits Janakalyan Yuba Club in Lahan, Siraha district said, "The program initiated by the government "Jaanta Awash" is not effective. We have found that the houses built under this project are not strong, as low quality materials have been used in construction." The rehabilitation measures are not adequate. There has been a lack of adequate information on risks, hazards and plans (disaster mitigation plans of local governments). According to Mr. Rajak – the Head of Madheshi Dalit Federation, to reduce the impact of climate change in the Madhesh province, it is especially important to protect the Chure Region, the hills on the northern edge of the plains that is facing chronic deforestation and degradation. At present, the 'Presidential Chure Conservation Program' is working for the protection of the Chure but experts are starting to question its beneficial impacts on Dalit communities. The future of Madhesh depends on Chure, it must be preserved to reduce the impact of the climate crisis in the region. Presidential Chure Conservation Program has also been under fire for the misuse of its funds and well as for their ineffective programs.

Implications of Climate Change on Economic, Social and Cultural Rights of Dalits: Climate change hurts the economic rights (e.g., rights to work, rights to receive a fair wage, and rights to a safe working condition) of Dalits. Their social rights, for example, right to social security, right to protection of the family, right to an adequate standard of living (freedom from hunger, access to clean water, adequate housing, and protection of property) have been affected too. Their cultural rights, for example, rights to education and the right to take part in cultural life, are also severely undermined due to climate change.

A DHRD and local resident of Bhagwanpur Rural municipality in Siraha district said that the Balan Khola overflows during the rainy season and no one gets a night's sleep, worried that they will be flooded overnight. Families here have stayed up all night and day at time on their rooftops. They couldn't harvest the paddy they planted. They get disconnected from the cities, from all the services including the use of toilets.

Similarly, according to a journalist in Janakpurdham in Madhesh, 50–60 households in a Musahar community were displaced from the site of a railway track. They did not have anywhere to go. The unplanned development has displaced many Dalits and they are forced to live in unsafe areas. During the monsoon, the Dalit community has to suffer much because the houses they live in are built of bamboo and mud, and houses are built in floodplains on riverbanks. Women and children are at even higher risk and live in fear every day.

Addressing Climate Change and Land Rights for Dalits

To protect Dalits from the ravages of climate change, several provisions of the Constitution should be implemented in both the letter and spirit. For example, the Seventh Amendment to the Lands Act has amended Section 52 to create a legal framework for implementing the constitutional obligation to provide land to landless Dalits. By this amendment, there is a requirement in the Lands Act to identify landless Dalits and provide a plot of land to them.

Unfortunately, these policies have not been developed in a way that addresses the dire impacts of climate change on Dalit communities. For example, no provision has been made to define landlessness and prescribe criteria to identify landless Dalits for the distribution of land under the Lands Act. The Amendment does not contain any statutory guidance in terms of a number of core aspects such as purpose (e.g., housing and sustainable livelihood through farming), quantum, nature, and quality of land to be provided to landless Dalits under this Act. It should also specifically address climate change vulnerabilities and fairness in disaster preparedness, relief, and recovery.

A study carried out by Amnesty International and its Partners in 2019 (AI Nepal, CSRC, Jury Nepal 2019) recommends the following to address shortcomings and ensure Dalit land policies are made with climate justice as a central value:

- Define landlessness and provide an explicit criterion to identify the landless Dalits eligible to receive land from the government.
- Make adequate legal arrangements to implement the provisions to identify landless Dalits and provide them with land for housing as well as resources for a sustainable livelihood as promised under the Constitution.
- Relocate Dalit settlements away from risky areas to safe places and allocate new lands for residence and livelihood. These new lands should be away from climate risk areas such as steep hillsides and riverbanks.
- When arranging housing for landless and homeless Dalit families, provide them with basic facilities like employment, education, health, roads, drinking water, electricity, and markets.

Although there is a legal obligation to complete the distribution of land within three years, there should be a provision such that the government is provided with an annual report on progress made.

Conclusions

This chapter addresses the issues of differential impacts of climate change, and climate injustice in relationship with Dalit communities, the most marginalized and vulnerable populations in Nepal.

It has demonstrated that landlessness and near landlessness among Dalits is a major risk factor for climate injustice and conversely, that equitable provision of land rights is a powerful means to buffer the effects of the climate crisis. This can help Dalit communities engage productively and from an empowered position in negotiating government policies and programs on climate mitigation, adaptation, and reparation for loss and damage. Secure land rights can also provide incentives for Dalits to invest in long-term land stewardship through the adoption of sustainable land management practices and the development of climate-resilient infrastructure. Developing and implementing these policies will require a transformation in the content and process of federal, regional, and local government policies that address the historical legacies of caste discrimination that has subjected Dalits to social, economic, and ecological marginalization.

More broadly, the relationships between caste discrimination and climate change offer some important insights into the emerging field of environmental and climate justice in Nepal and elsewhere in the Global South. In particular, the study emphasizes that in the discourse of climate justice, the climate crisis should be recognized as a social and political problem as well as an environmental one. Dalits being at the lowest strata from socio-political and economic perspectives are most affected by climatic disasters, and to ensure climate and environmental justice to these most disadvantaged communities in Nepal, their sufferings aggravated by differential impacts, landlessness, and their fragile environments should be rightfully addressed.

Finally, the Samata Foundation's report (Pariyar, 2022) on climate change and Dalit communities is the first of its kind but only provides a broad overview of the topic. As such, more detailed studies should be done to examine the specific conditions of landless and near-landless Dalits (as well as other marginalized populations) and their needs and priorities related to climate change.

References

AI Nepal, CSRC, Jury Nepal (2019). Nepal: Land for Landless Peasants – Comments and Recommendations on Amendments to the Land Act 1964. Amnesty International, Community Self Reliance Center Nepal, Jury Nepal.

CBS (2013). National Sample Census of Agriculture Nepal 2011–2012 National Report. Central Bureau of Statistics, National Planning Commission, Government of Nepal, Kathmandu Nepal, December 2013.

Dhakal, S. (2011). *Land Tenure and Agrarian Reforms in Nepal*. Kathmandu, Nepal: Community Self-Reliance Center.

Dhaulakoti (2022). Sukumbasi ka Naam ma 30 Barsha ma 13 Aayog (13 commissions in 30 years in the name of Squatters) Prakash Dhaulakoti, Kantipur Daily, December 3, 2022.

Dulal and BK (2019). Bishwa Bhakta Dulal (Aahuti) and Ganesh B.K., *Dalit Bhumihinta ra Sambidhan Karyanwayan ko Prashna* (*Dalit* Landlessness and Question of Constitution Execution). Community Self Reliance Center (CSRC).

Giri, B. (2009). The Bonded Labour System in Nepal: Perspectives of Haliya and Kamaiya Child Workers. *Journal of Asian and African Studies* 44(6), 599–623.

IIED (2020). Helping Indigenous Communities Secure Land Rights in Nepal, 18 December 2020. https://www.iied.org/helping-indigenous-communities-secure-land-rights-nepal.

Nepali, P. B. (2008). Access to Land Resource: Dalits and their Livelihood Insecurity. Nepal: Transition to Transformation. Kathmandu: HNRSC-Kathmandu University and SAs RCO NCCR North-South, 163–184.

NHDR (2020). *Nepal Human Development Report-Beyond Graduation: Productive Transformation and Prosperity. National Planning Commission.* Government of Nepal and United Nations Development Programme, Kathmandu, Nepal.

Panday, U. S., Chhatkuli, R. R., Joshi, J. R., Deuja, J., Antonio, D., & Enemark, S. (2021). Securing Land Rights for All through Fit-for-Purpose Land Administration Approach: The Case of Nepal. *Land* 10, 744. https://doi.org/10.3390/land10070744.

Pariyar (2022). Madan Pariyar. Study on Causes of Climate Vulnerability, Conflict Dynamics and Existing Local Adaptive Capacities of *Dalit* Communities in Nepal, Samata Foundation, December 2022.

Ranganathan, M. (2022). Caste, Racialization, and the Making of Environmental Unfreedoms in Urban India. *Ethnic and Racial Studies* 45(2), 257–277.

Sharma, M. (2022). Caste, Environment Justice, and Intersectionality of Dalit–Black Ecologies. *Environment and Society* 13(1), 78–97.

8 Environmental Justice and Pesticides

Kishor Atreya, Kanchan Kattel, Anisha Sapkota, and Hom Nath Gartaula

Introduction

Agricultural intensification in Nepal has led to a high application of pesticides, which pollute the land, water, and soil and degrade the health of agroecosystems and peasants (Acharya et al. 2022; Bhandari et al. 2020, 2019; Atreya et al. 2011) (see Figure 8.1). Pesticide consumption has increased over ten-folds since 2000, and the import trend has been increasing over the years. The national average pesticide use is estimated to be around 400 grams of active ingredient per hectare

Figure 8.1 Pesticide-laden beans ready for transportation to market.

Photo credit: Authors.

DOI: 10.4324/9781003371175-10
This chapter has been made available under a CC-BY-NC-ND 4.0 license.

(PPD 2018). However, in some areas, extremely high use of pesticides has been reported for vegetables (Bhandari et al. 2018). As of July 2022, a total of 24 chemical pesticides are banned for use and 165 are registered for agricultural use in Nepal (PQPMC 2023). Of the registered pesticides, approximately 34% are insecticides, followed by 26% fungicides, 9% biopesticides, 8% herbicides, and others (acaricides, nematicides, bactericides, rodenticides, molluscicides, and herbal pesticides). Almost half of these registered pesticides belong to Class II (moderately hazardous) of the WHO classification (WHO 2020). A large amount of pesticide is used in vegetable crops (>90%), followed by cereal crops (Aryal et al. 2021; Ghimire and GC 2018). The rate of pesticide consumption per unit of land is still lower in Nepal than in other parts of the world; however, this is changing.

While there are some positive benefits of pest control such as increased income and improved livelihoods, pesticide use also brings several environmental and health-related impacts and economic burdens (Atreya et al. 2012). For example, a significant positive association has been established between a history of pesticide use and multiple chronic health problems (Atreya et al. 2020; de-Assis et al. 2020). Furthermore, Nepali farmers rarely show complete adherence to safety measures during pesticide handling and application (Atreya et al. 2022; Bhandari et al. 2018).

Due to the well-observed negative consequences of pesticide use, the Nepali government has been promoting alternative pest control measures such as integrated pest management (IPM) and biopesticides, which are less costly and more environmentally friendly (Paudel et al. 2020), but these alternatives have not been well adopted by farmers for numerous reasons, such as lowered yields and the inconvenience on large and/or remote farms. For such reasons, chemical pesticides continue to be the most preferred plant protection technique, resulting in various environmental, health, and social burdens.

The health and pollution risks to humans and the environment and the associated burdens are disproportionately distributed between particular social groups, resulting in environmental injustice. For example, Atreya et al. (2013) demonstrated that small-scale households were deprived of the benefits of pesticide use and also incurred their greatest health and environmental burden. Similarly, Atreya (2007) and Garcia (2003) found a significant difference in pesticide use and safety measures between men and women. Environmental justice in pesticide use can only be achieved if all social groups have equal and fair access to information, income benefits, and livelihood opportunities, and do not face inequitable risks to their health and the environment.

In Nepal, several studies (Atreya et al. 2022 and references therein) have considered household-level use of pesticides and pesticide handling practices, but very few of them have addressed the distribution of the health and environmental burden across society (but see, Atreya et al., 2013). Most studies have assumed that "farmers" are homogeneous units in their analysis and interpretation; however, the benefits and burdens of pesticide use, health impacts, and environmental pollution can differ according to caste and gender, among other differences. The means and statistics from such studies only denote the population as a whole and therefore the unequal distributions of exposure, risk and opportunities to the vulnerable sub-groups within the population generally remain unrepresented (Gochfeld and Burger 2011). Limited studies have examined pesticide use and social marginalization in terms of the distribution of associated benefits and burdens, the recognition of opportunities and problems, and the participation in its management.

The objective of this paper is to assess whether the risk of agricultural chemical pesticide use differs according to factors of social marginalization, mainly by gender and caste. In turn, this will help develop a fundamental understanding of environmental justice regarding pesticide use in Nepal. We analyze three dimensions (Isgren and Andersson 2021) of environmental injustice: (i) distribution-how goods (benefits) and bads (burdens) are distributed between groups?

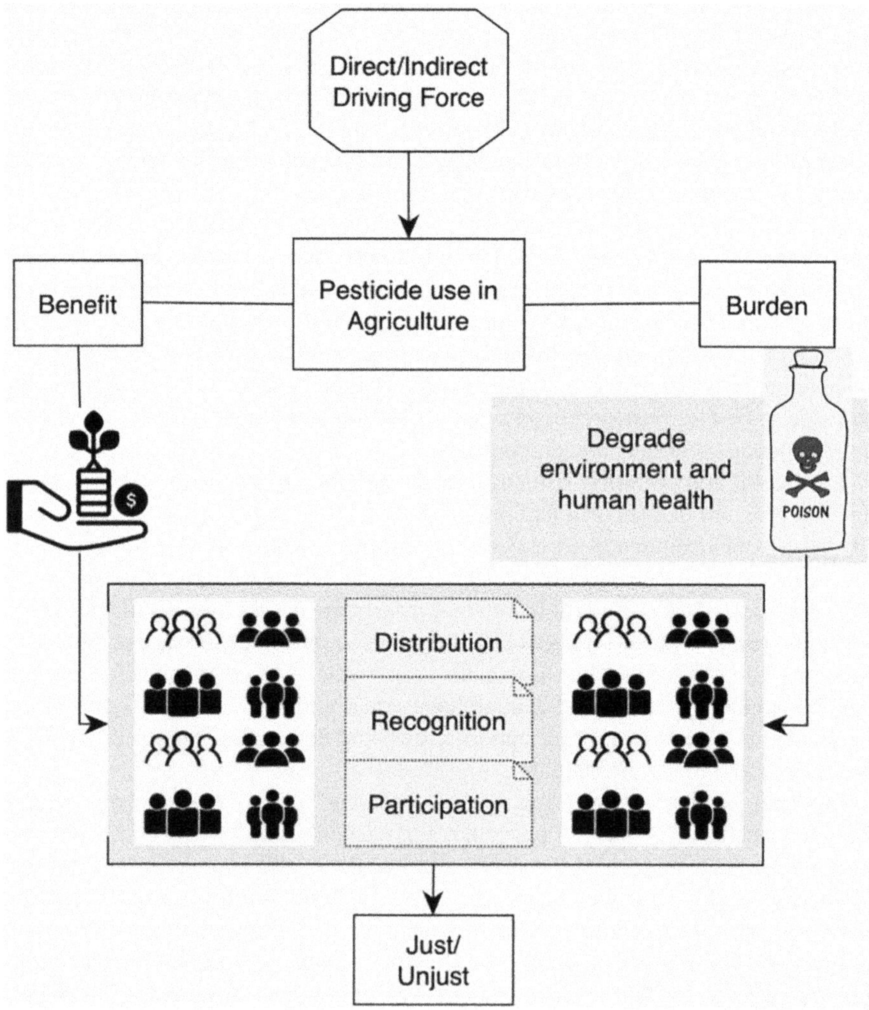

Figure 8.2 Revisiting pesticide use from an environmental justice lens.

Source: Authors' visualization.

(ii) recognition – how good(s) and bad(s) are perceived between groups? and (iii) participation – how do groups participate in decision-making regarding managing goods and bads? (See Figure 8.2).

Study Area

This chapter is based on case studies of Mahadevsthan farmers, a former Village Development Committee (rural municipality) in Kavre district, in the middle mountain area, east of Kathmandu. Numerous farmers in Mahadevsthan practice commercial vegetable production for Kathmandu markets and have used chemical pesticides for several decades. Mahadevsthan farmers generally belong to three ethnic groups: Brahmin-Chhetri-Thakuri or BCT (so-called "privileged caste group"), Janajati (indigenous caste group), and Dalits (the so-called "oppressed caste group" who have been historically marginalized as "untouchables"). To study the differences between groups from the perspective of justice, we included farmers belonging to these three caste groups. We also included male and female focus group discussions from all the caste groups to explore gendered differences in perspectives, practice, and experience.

Methods

We conducted 11 focus group discussions or FGDs (2 Dalit groups, 2 Danuwar Janajati groups, 1 Tamang Janajati group, 3 (1 BCT, and 2 mixed-castes) males only, and 3 (1 BCT and 2 mixed castes) females only; and 12 key informant interviews with different key stakeholders, farmers trained in IPM, high-pesticide users, and commercial farmers. Interviews and discussions were facilitated by a team member and discussion notes were written by another, and interviews were also recorded. Prior informed consent was obtained verbally from each of the participants at the time of recording. Interview and discussion summaries were written for all 23 recordings. The entire conversation was not transcribed, but the relevant information and important quotes from the participants were translated into English.

Results

The burdens and benefits of chemical pesticide use in agriculture were disproportionately distributed by gender and caste. See Tables 8.1 and 8.2.

Pesticide Use and Gender

Male members of the family are generally in charge of pesticide handling, spraying, and related decision-making when males live at home, which largely depends on the type of family and the occupation of the husband. Although generally men apply pesticides, the participation of women in pesticide handling is not negligible, and women are also at risk of exposure through handling, application, and indirectly during farm work right after pesticide application when residue levels in farms are high enough to cause acute health problems. In the case of male absenteeism (often

Table 8.1 Disproportionate distribution of the effect of pesticide use between men and women

Dimensions	*Men*	*Women*
Distribution of benefits and burdens (e.g., vegetable income, and risk of exposure)	Responsible for the handling of pesticides in most households, hence more exposure.	Responsible for pesticide handling in most households with male absenteeism.
Recognition of health risk and safety	Health risk is less acknowledged; less adherence to safety precautions (associated with the definition of masculinity).Health risk is part of "agricultural life."	Higher recognition of health burdens, but limited adherence to safety measures.
Participation in management (e.g., Jholmol and IPM)	Domination in decision-making and management; however, intra-household interplay has a role.	Limited access to training and awareness programs, thus less involvement.

due to labor migration out of the country), women are the only applicators. In the case of hired labor for pesticide applications, almost all farmers preferred men.

Although acute health symptoms are observed in both men and women, interviewees believed women to be more vulnerable to certain life situations, mainly during sensitive reproductive stages such as pregnancy, postpartum, or menstruation cycles. The increased vulnerability of women to pesticide use was also attributed to the clothing style of women from Nepali villages. "Many of the women here do not wear underwear and are more vulnerable to getting sick from pesticides. Many women in the area are suffering from cervical cancer and related diseases," said a participant belonging to the BCT caste group and working as a community health volunteer, while another BCT participant in the next FGD also agreed. Farmers in lowland areas (*besi*, where pesticide use is comparatively higher compared to higher areas) reported an increase in cervical cancer cases in the community due to the excessive and frequent use of pesticides.

Many male participants in FGDs reflected that they were strong enough to digest the pesticides and that it would not affect them much compared to women participants who said it inevitably affected their health. A participant in male-only FGD said, "I have tasted Dithane *[mancozeb, a fungicide]* and nothing happened to me as a result." Likewise, an IPM-trained male farmer during KII stated that he had been exposed to Metacid (methyl parathion, a highly toxic insecticide, banned in Nepal since 2007) and felt severe burns in the stomach, and now he has completely abandoned the use of chemical pesticides due to health consciousness.

Regarding the distribution of training opportunities, women's access to IPM training varies depending on their ethnicity and place of residence. Many of the BCT women had attended IPM training; however, women from the Danuwar and Dalit groups had rarely received such opportunities. Even among BCT women trained with IPM, knowledge of the pesticide toxicity labels on the containers, for example, was very low (10%). This suggests a lower knowledge of women and even fewer opportunities for women of marginalized caste groups.

Table 8.2 Disproportionate distribution of the effect of pesticide use between ethnicities

	Brahmin Chhetri Thakuri	*Janajati*	*Dalits*
Distribution of benefits and burdens (e.g., vegetable income and risk of exposure)	Significant improvement in livelihood through pesticide use in commercial vegetable farming (high landholding and increased production); and less risk to health because more training and safety advice is received.	Tapping opportunities to improve livelihoods to some extent (less landholding, so renting BCT's land for production); high health risk due to inadequate trainings; have lost some parts of their income share from ecosystem services (e.g., fishing).	Less benefited by the opportunity to increase the yield using pesticides (marginal landholding and rental access); Less direct health risk, as pesticides are not used much in subsistence farming.
Recognition of health risk and safety	Good recognition and analysis of pesticide-related benefits and burdens; Health safety is still less prioritized after adequate participation in awareness programs.	Very poor understanding and recognition of benefits and burdens.	Lowest level of understanding and recognition of benefits and burdens.
Participation in management (e.g., Jholmol and IPM)	Increased participation in training and cooperative initiatives related to IPM and other ecological alternatives, although adoption is not satisfactory due to perceived drudgery.	Lesser participation in IPM trainings, local cooperative initiatives related to IPM, and other alternatives	Least participation due to poverty, neglect, and systematic marginalization.

Male domination was observed in pesticide decision-making and management, especially in the BCT households. For example, an IPM-trained BCT female said that her husband sprayed chemical pesticides on her farm without her approval because neighbors applied chemicals. Likewise, a Dalit woman stated that her husband takes care of all activities related to pesticides, whereas another Dalit woman opposed it, and stated that it varies. But in the BCT women FGD, many participants said that intra-household interplay (e.g., male absenteeism) determines the primary applicator and subsequent pesticide-related decision-making process.

Pesticide Use and Ethnicity

In the study area, Dalits were less engaged in commercial agriculture and therefore less involved in pesticide use due to their small landholdings and the relatively high ownership of more marginal land. However, the Janajati caste groups are more involved in commercial farming, both on their own and leased land, resulting in more pesticide exposure and the possibility of experiencing higher burdens.

Both Dalits and Janajatis receive fewer opportunities to receive knowledge, such as training, and benefits, such as government subsidies and other opportunities, suggesting greater risk due to their limited knowledge and awareness of pesticide toxicity and safety. Farmers in these caste groups also wrongly perceived minimal health risks from unsafe pesticide use and pesticide pollution. According to an agricultural veterinary personnel (agrovet),

> The indigenous minority of Danuwar often wants quick results and demands more toxic pesticides (*kada*). Pesticide-related programs, such as IPM training and awareness-raising programs, are more concentrated in the lowlands (*besi*), where Danuwars are less, and thus have not reached out properly to these farmers.

In male-only FGD, a Tamang farmer said, “We use a dose a little higher than recommended by the agrovet personnel” and also reported that the doctor associated his health problem with his previous exposure to pesticides: “When I went for a health check-up for neural and blood circulation problems, the doctor related it to the use of pesticides.” On the contrary, some farmers believe that agrovets have a tendency to sell more pesticides to uninformed farmers and that the amount of pesticides given for the same problem is different for IPM-trained and untrained farmers. In the FGD with Tamang group, many participants stated that: “When trained farmers go to buy pesticides, agrovet gives them only one pesticide; otherwise, agrovet asks to mix various pesticides and sell more than one pesticide. You know it is just like buying drugs from private medicals.”

BCT farmers perceive and recognize the threat of chemical pesticides due to their greater participation in related training and awareness-raising programs. However, they still use minimal PPE due to concerns about convenience and discomfort, which results in increased exposure and health burdens. In comparison, BCT farmers trained in IPM use PPE more effectively and have a higher risk perception, resulting in the lowest health risk. Unfortunately, compared to other production methods, IPM techniques are less widely used due to their lower crop yield, high labor cost and labor shortage, more work, and low market premiums.

Regarding economic burden, especially for the Danuwars, their primary and traditional income opportunity, namely fishing, has been compromised due to pesticide pollution in irrigated rice fields and rivers, but it is less recognized and realized.

> In the past, my father and my elder brother would practice fishing using *“chhitri or thitri”* [an almost cylindrical shaped trap made up of Indian gooseberry twigs to trap small fish from rice fields and rivers] on other people’s land and sometimes even sell them to *Tamaghat and Panchkhal* [adjacent market outlets].

“Mostly the harvested fishes were consumed at homes for better nutrition, but in the case of surplus, a household used to sell on average 30 kg in a paddy growing season." When asked how many households currently use *çhhitri*, they replied saying only two out of hundreds. This reduction in fishing is mainly due to the use of pesticides and chemical pollution. "It is due to the toxic pesticide, why not? Otherwise, there used to be so many fish and snakes in the past. Now people cultivate the land and use a lot of pesticides. Such toxic chemicals have killed them," said a key informant.

Discussion

Historically, national and subnational policies and cultural norms have favored men, higher caste and class groups, and other elite sections of society in agriculture and other sectors, and these disparities have continued with increased globalization and trade liberalization. Cultural discrimination, economic exploitation, social exclusion, and political oppression issues have often been critiqued, yet still exist in Nepal (Gurung 2009). This exclusion has led to systematic marginalization of women and ethnic minorities, and the impacts are observed in, but not limited to, resource use and societal benefits. Studying pesticide use and associated impacts shows that this marginalization leads to disproportionate impacts, both positive and negative. In our research, we found that men and women have different levels of participation in making decisions, and this difference becomes even greater based on the ethnic background of farmers. This means that women belonging to certain groups that have historically been treated unfairly have the least say in handling issues related to pesticides.

Risk perception and safety precautions, two important ways to minimize the health burden of pesticide use, are distributed differently based on gender and ethnicity. Knowledge appears to be lacking in ethnic groups of marginalized strata. Likewise, participation in risk management and control over decision-making is lacking both among women and among these same caste groups. In other words, the access to knowledge-generating opportunities is mostly biased toward privileged caste groups, and the participation in decision-making is also biased towards households from these groups and men from such households. This is perhaps a result of the general elite capture of opportunities and the *bhagbanda* (benefit sharing) among influential people in the society (Acharya et al. 2022). Men generally are responsible for handling pesticides, but even with better knowledge and awareness of associated health risks, fewer of them adhered to safety behaviors when using pesticides, in line with the findings of previous studies (Wang et al. 2017). This exclusion in participation in knowledge generation and management further impacts the recognition of problems and solutions, burdens, and benefits among the different groups we studied.

In addition, livelihood degradation is one of the gravest outcomes of pesticide consumption, in which privileged farmers receive better opportunities to earn a living from pesticide use, while it diminishes the livelihood opportunities for those belonging to more marginalized caste groups, such as the Danuwars in Kavre. This latter caste group traditionally relied on fishing activities, which meant that they would have free access to fish even on the land of other privileged people, since they themselves had limited landholdings. However, this income and nutrition opportunity no longer exists for Danuwars, as pesticide pollution has led to a massive decline in fish population, leading to the only remaining option where they could rent the land from farmers of privileged caste groups with relatively large landholding, further generating passive income for those families and limited profit for the Danuwars who actually farm the land.

Active participation of the privileged community (Brahmin/Chhetri) in IPM practices like *jholmol* initially gained a lot of attention. Their proactive involvement was a positive change. However, a problem arose as they started doing less and less farming. This happened because many of their children went to work/study in other countries, which reduced the amount of work they could do on the farm. These migrant family members also encouraged their parents to do fewer farming (key informants). The Janajati community faced a different challenge. They did not learn as much during the training phase, so they adopted *jholmol* more slowly. Similarly, the Dalit community has also faced difficulties using *jholmol*. The Dalits had the least access to *jholmol* because other communities and programs ignored them, and they also faced economic difficulties. They were systematically neglected in participation to learn about *jholmol* during the training phase, and now they are moving away from traditional farming and looking for work in off-farm jobs. This complicated situation shows how different factors such as money, culture, and knowledge affect how these different communities are able to use eco-friendly farming methods.

Finally, there are some external elements to consider, such as the profit-making private sector which directly or indirectly influences the use of pesticides, the use of safety equipment, the dose, and, in turn, the exposure to pesticides. With poor extension services in these communities, privately run agrovets take over the role of government extension services, enjoy a monopoly on the market, and thereby control the agri-food system. This has a significant impact on both the environment and the health of people. Similarly, the introduction of a holistic and system-thinking approach in local municipalities is suggested for the management of pesticide injustice. For example, local municipalities at present have three different functional sections: health, agriculture, and veterinary; and coordination and interaction between them are weak because of overlaps and inconsistencies in the allocated functions. Therefore, we suggest either establishing good coordination or merging these functional units for overall monitoring of pesticide use and evaluation of its benefits and burdens to society. Otherwise, the unregulated introduction of seeds, chemicals, and technologies will threaten local agrobiodiversity and the rich genetic agricultural and environmental resources that support livelihood, health, culture, and other aspects of farmer life, especially the poor and marginalized, and further intensify environmental injustice.

Conclusions

Both men and women are exposed to pesticides, but women have a higher health risk due to their exposure to pesticides during vulnerable life stages. Men of privileged castes participate in most training and awareness-raising programs, are more aware of the danger of pesticides, but adhere less to the safety precautions in pesticide handling. Women are less aware but adopt higher safety measures because they perceive a greater danger of exposure to pesticides.

Most of the opportunities, including IPM training and other support systems, are highly received by BCT, but BCT lands are often cultivated by indigenous groups, and therefore are at greater risk of exposure. Dalits, on the other hand, are less exposed to pesticides because they have smaller farms. Danuwar, an indigenous group, has found that pesticides have almost completely destroyed local fishing, their traditional subsistence method.

These dynamics are complex; therefore, ensuring a just interaction between pesticide use, the environment, and people's health is difficult to achieve with traditional extension and awareness approaches. We recommend more holistic and inclusive and accessible training, education, and awareness programs by designing and implementing interventions tailored to pesticide use in agriculture and acknowledging the burdens of negative effects of pesticide use in the policy decision (e.g., social benefit-cost analysis), especially at the time of pesticide registration and approval. These programs must address the ongoing legacy of caste and gender discrimination in access to information and resources.

Scientific studies that analyze group differences in pesticide exposure and impacts and those that involve clinical studies are limited in Nepal, highlighting the need for more such clinical studies to better understand the health burden of pesticides. In general, internalizing the multidisciplinary aspects of pesticide-related issues and focusing on the "agriculture-health" nexus should be the top priority to ensure a just agriculture system for both people and the environment.

Acknowledgment

This paper is the result of a research grant to the lead author. We thank the financial support of the Higher Education Reform Project, Tribhuvan University, Nepal (HERP-DLI-7B). The valuable time of the participants provided in the focus group discussions and key informants is well acknowledged.

References

Acharya, K. K., Dhungana, R. K., & Guragain, H. P. (2022). The position of marginalized groups in the elite captured local level planning process in Nepal. *Nepal Public Policy Review* 2: 1–26.

Aryal, S., Dangi, N., & Simkhada, R. (2020). Chapter 13 Trends in Pesticide Use in Different Agricultural Commodities and Residues in Nepal in S. Shyaula, G. Bajracharya, G. K.C., S. Shakya & D. Subba (Eds.) *Comprehensive Insights in Vegetables of Nepal.* Lalitpur, Nepal: Nepal Academy of Science and Technology, 461–483.

Atreya, K. (2007). Pesticide use knowledge and practices: A gender differences in Nepal. *Environmental Research, 104*(2), 305–311. https://doi.org/10.1016/j.envres.2007.01.001

Atreya, K., Johnsen, F. H., & Sitaula, B. K. (2012). Health and environmental costs of pesticide use in vegetable farming in Nepal. *Environment, Development and Sustainability, 14*(4), 477–493. https://doi.org/10.1007/s10668-011-9334-4

Atreya, K., Kattel, K., Pandit, S., Chaudhari, P., & Sipkhan, P. (2022). Understanding farmers' knowledge, attitudes and practices of pesticide use in Nepal: Synthesis of a systematic literature review. *Archives of Agriculture and Environmental Science, 7*(2), 278–287. https://doi.org/10.26832/24566632.2022.0702018

Atreya, K., Rijal, C., & Neupane, N. (2020). Pesticide use in agriculture and chronic health conditions: A survey-based cross-sectional study in Nepal. *Archives of Agricultural and Environmental Science, 5*(4), 489–497.

Atreya, K., Sitaula, B. K., & Bajracharya, R. M. (2013). Distribution of health costs of pesticide use by household economy. *Environment, Development and Sustainability, 15*(3), 827–839. https://doi.org/10.1007/s10668-012-9414-0

Atreya, K., Sitaula, B. K., Johnsen, F. H., & Bajracharya, R. M. (2011). Continuing issues in the limitations of pesticide use in developing countries. *Journal of Agricultural and Environmental Ethics, 24*(1), 49–62. https://doi.org/10.1007/s10806-010-9243-9

Bhandari, G., Atreya, K., Scheepers, P. T. J., & Geissen, V. (2020). Concentration and distribution of pesticide residues in soil: Non-dietary human health risk assessment. *Chemosphere, 253*, 126594. https://doi.org/10.1016/j.chemosphere.2020.126594

Bhandari, G., Atreya, K., Yang, X., Fan, L., & Geissen, V. (2018). Factors affecting pesticide safety behaviour: The perceptions of Nepalese farmers and retailers. *Science of the Total Environment, 631–632*, 1560–1571. https://doi.org/10.1016/j.scitotenv.2018.03.144

Bhandari, G., Zomer, P., Atreya, K., Mol, H. G. J., Yang, X., & Geissen, V. (2019). Pesticide residues in Nepalese vegetables and potential health risks. *Environmental Research, 172*(March), 511–521. https://doi.org/10.1016/j.envres.2019.03.002

de-Assis, M. P., Barcella, R. C., Padilha, J. C., Pohl, H. H., & Krug, S. B. F. (2020). Health problems in agricultural workers occupationally exposed to pesticides. *Revista Brasileira de Medicina do Trabalho, 18*(3), 352–363. https://www.ncbi.nlm.nih.gov/pmc/articles/PMC7879472/

Garcia, A. M. (2003). Pesticide exposure and women's health. *American Journal of Industrial Medicine, 44*, 584–894.

Ghimire, K., & GC, A. (2018). Trend of Pesticide Use in Nepal. *Journal of the Plant Protection Society 5*: 32–42.

Gochfeld, M., & Burger, J. (2011). Disproportionate exposures in environmental justice and other populations: the importance of outliers. *American Journal of Public Health, 101*(S1), S53–S63.

Gurung, O. (2009). Social inclusion: Policies and practices in Nepal. *Occasional Papers in Sociology and Anthropology, 11*, 1–15.

Isgren, E., & Andersson, E. (2021). An environmental justice perspective on smallholder pesticide use in Sub-Saharan Africa. *Journal of Environment and Development, 30*(1), 68–97. https://doi.org/10.1177/1070496520974407

Paudel, S., Sah, L. P., Devkota, M., Poudyal, V., Prasad, P. V. V., & Reyes, M. R. (2020). Conservation agriculture and integrated pest management practices improve yield and income while reducing labor, pests, diseases and chemical pesticide use in smallholder vegetable farms in Nepal. *Sustainability, 12*(16), 6418. https://doi.org/10.3390/SU12166418

PPD. (2018). *Study on national pesticide consumption statistics in Nepal*. Ministry of Agriculture, Government of Nepal. Lalitpur, Nepal.

PQPMC. (2023). *List of registered pesticides and pesticide consumption data – 2079* [in Nepali]. Plant Quarantine and Pesticide Management Centre (PQPMC), Ministry of Agriculture and Livestock Development, Government of Nepal. Kathmandu, Nepal.

Wang, W., Jin, J., He, R., & Gong, H. (2017). Gender differences in pesticide use knowledge, risk awareness and practices in Chinese farmers. *Science of the Total Environment, 590–591*, 22–28. https://doi.org/10.1016/j.scitotenv.2017.03.053

World Health Organization of the United Nations (WHO). (2020). *WHO recommended classification of pesticides by hazard and guidelines to classification*, 2019 edition. Geneva.

9 From Red to Green to Grey Hills

Reflections on the Four-Decade-Long Journey of Community Forestry and Environmental Justice in Nepal

Sunita Chaudhary

Introduction

Nepal is a pioneer in community forestry which has set an example of a significant increase in forest cover, decentralisation, and local people's participation in forest management (Shahi et al., 2022). Through the 1970s, the country suffered from high deforestation and landslides – termed as the "Era of Himalaya degradation" by Erik Eckholm in his book *Losing Ground* (1976). Community forestry first came into existence in the late 1970s with an aim to halt deforestation, expand and conserve forest cover, and support local people's subsistence livelihoods (Smith et al., 2023; Chaudhary, 2017). Over the last four decades, community forestry has achieved significant progress in pursuing dual conservation and development goals. It has been recognised for increasing forest cover, fulfilling forest product and livelihood needs, addressing inequality, strengthening women leadership, and more recently tackling climate change, and hence achieving sustainable development agenda (Laudari et al., 2024).

Today, there are more than 22,000 community forest users' groups (CFUGs) including approximately 2,900,000 households who are managing about 2,200,000 hectares of forests across the country (Forest Research and Training Centre [FRTC, 2022]). Nepal's community forestry programme has gained attention from around the world for greening the mountains, conserving soils, decentralising the forest management system, and supporting subsistence livelihoods, and climate change adaptation.

Although pioneering and popular, community forestry has gained substantial criticism as well (Giri, 2022; Ojha, 2006; Malla, 2000; Malla 2001). Community forestry has been criticised for the technocratic domination of forest science, and marginalisation of local voices in governance (Ojha, 2006). For instance, requiring inventory-based management plans as a pre-requisite for community forest management allows the forest technicians to control communities and influence forest governance (Baral et al., 2019). Community forestry has also been criticised for entrenched inequity in terms of access to benefits. Limited participation and recognition in the community forestry system are other critiques of the system (Chaudhary et al., 2018; Thoms, 2008). Lack of recognition of marginalised communities especially in the public sphere and their limited participation

DOI: 10.4324/9781003371175-11
This chapter has been made available under a CC-BY-NC-ND 4.0 license.

in decision-making is widespread across the community forests in Nepal (Sunam & McCarthy, 2010). The marginalised communities, including women, Dalits, Janajati, and poor people, often bear the costs of improved forest conditions with strict forest protection and limited access to forest resources (Thoms, 2008). Recently, community forestry has been criticised for dysfunctional CFUGs with limited local participation (Laudari, 2024; Ojha, 2023). Increased forest fires have been linked to inequity issues, declined people's dependence on forests, and neglecting local people's priorities in community forestry (Paudel et al., 2021).

Reflecting on the history of community forestry in Nepal, this chapter analyses the changing paradigms of community forests over the last four decades. The reflections and discussions are presented in three different phases:

1. Red Hills (1950s–1980s): "Himalayan crisis" dilemma in pre-community forestry era.
2. Green Hills (1990s–early 2000s): Community forestry era with increased forest cover and functional CFUGs but marred by systemic injustices.
3. Grey Hills (2000s–present): Increased and intense forest fires with heavy smoke across the hills of Nepal and dysfunctional CFUGs.

In each of the phases, the social and ecological aspects of community forestry are discussed, together with their equity and justice implications (Figure 9.1).

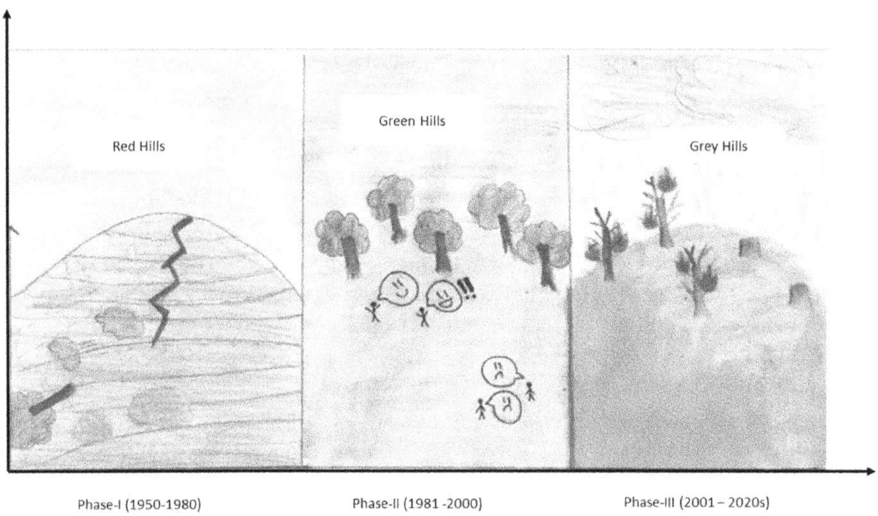

Figure 9.1 The history of community forestry in three different eras.

Credit: Lavanya Chaudhary.

Changing Paradigms and Associated Justice Issues of Community Forestry Red Hills (1950s–1980s)

This phase has been termed as "Red Hills" because of the high exposure of soils leading to erosion, denuded hills, landslides, and flooding across the hills of Nepal (Eckholm, 1984). The crisis started before the 1950s when the Rana government converted forests to agricultural land starting in the late 1800s. After the end of the Rana regime in 1951, the royal Shah government continued to clear forests to initiate nationwide development of roads, hospitals, industries, and schools (Malla, 2000). A large swath of hill forests were cleared to convert into agricultural land to generate tax for the government, while the Terai forests were cleared to meet the timber demands of India (Blaikie et al., 2002). Between 1947 and 1980, forest coverage declined from 57% to 23% (Myers, 1986). The degradation of hills created intense landslides, flooding, and decreased soil fertility by the 1980s (Gilmour & Fisher, 1991).

Justice and Equity Issues during the Red Hills Phase

The nationalisation of forests in 1957 limited access to forest benefits. Indigenous, Dalit, and other marginalised communities who had been traditionally managing and using the forest resources had limited or no legal rights to the forests. Control remained with the landlords or local elites or government officials. The Land Reform Act (1958) also contributed to the degradation of forests (Malla & Hobley, 2022). The *Forest Act 1961* and *Forest Protection – Special Provisional Act 1967* provided strong protection for forests, further restricting local access to forest resources. Fines and penalties were imposed on the marginalised communities with long histories of forest stewardship trying to access forest resources for their subsistence (Malla, 2000). The local elites grew trees on their private land to meet their forest needs, while the poor farmers had to depend on the local elites and landlords for forest products. Tight alliance between the local elites and local government-controlled access (Malla & Hobley, 2022).

The "Theory of Himalaya Degradation" unfairly blamed deforestation on the increasing population growth of the local mountain communities and their higher dependency on forest resources such as fuelwood and fodder for subsistence living (Ives & Messerli, 1989). The local people were blamed as the major driver of the "Red Hills" degradation (Malla & Hobley, 2022; Ojha et al., 2014). However, the theory over-simplified the complexity of land degradation as an environmental issue and disregarded the underlying sociopolitical factors (Ives & Messerli, 1989). More importantly, the theory blamed the local mountain communities, especially the poor and the marginalised communities, for degrading the land – exactly those who themselves were the victims of the degradation (Blaikie, 1985). Instead, scholars such as Blaikie (1985) pointed to systemic inequities produced by the state and social elites for pushing these marginalised peoples onto marginal lands and into unsustainable conditions. In fact, by reducing the availability of resources and increasing the time required for collecting forest products, forest degradation undermines the income, health, nutrition, and food security of local communities (Baland et al., 2009).

Green Hills (1990s–Early 2000s)

In the late 1980s following over two decades of nationalised forests, the new community forestry policy significantly handed over the national forests to local communities for conservation, management, and subsistence use (Malla, 2000). Local communities once identified as the drivers of the Himalayan environmental crisis became the agent of change for addressing the crisis. The concept of "people's participation" in forest management was introduced in the National Forestry Plan 1976 (Gautam et al., 2004) and the Forest Act 1993 legitimised the community-forest user groups (CFUGs) as a self-governing institution with rights to access, manage, and sustainably use national forests (Thwaites et al., 2017).

As community forestry gained momentum, the highly denuded hills became green all around the mid-hills of Nepal. Outmigration from the hills to urban centres and global diasporas also contributed to reduced demand for forest resources and increased reforestation (Oldekop et al., 2018; Hobley, 2012). By 2017, the forest cover significantly increased to 44.7% of the total area of the country covering 147,734 square kilometres (Department of Forest Research and Survey [DFRS, 2018]). Over the 20-years, community forestry received significant global attention for not only greening the mountains but also meeting the basic forest needs of the local population, which are seen as crucial to subsistence living and rural economies in Nepal.

Women's participation has been reported to improve forest conditions and governance of community forest (Agarwal, 2009) and community forestry has benefited from and strengthened women's leadership (RECOFTC, 2022; Hobley, 2012). It has improved the welfare of disadvantaged communities in rural Nepal (Maharjan, 2009). This is the result of struggles for justice by the disadvantaged communities who have been building their capacities to break structural barriers and raise their voice in decision-making (Hobley, 2012).

Justice and Equity Issues during the Green Hills Phase

Yet, justice and equity problems plagued the new community forests. Despite the special provisions for marginalised communities, elites captured many of the benefits (Yadav et al., 2015). Special rules established in 2008 and 2014 required 50% women representatives, either the chairperson or the secretary of the committee to be a woman, and proportionate representation from the poor, lower-caste, and ethnic groups in management committees (Government of Nepal [GoN, 2014]). However, in reality, access to forest ecosystem services was uneven (Pokharel & Tiwari, 2013). Marginalised groups and women, especially women from the marginalised communities, faced barriers to participation (Chaudhary et al., 2018). Lack of recognition in interpersonal and public platforms exacerbated powerlessness of marginalised communities leading to either forced or passive participation (Sunam & McCarthy, 2010). Moreover, instances of corruption and mismanagement persisted, where leaders or management committee members abused positions for personal gains (Basnyat et al., 2023). This was especially evident in Terai community forests due to fertile soil, high-value timber, and dense population. Local political leaders and landholders dominated the executive committees and gained more benefits (Bhusal et al., 2018).

As a result of these inequities, the Green Hills made possible through community forestry remained inequitably accessible to most needy populations, including those who were their traditional stewards. While new institutional forms of forest governance helped improve environmental conditions, they have not been able to surmount long-standing gender, caste, and other social systems of discrimination. In response, local and national organisations by these marginalised populations are strongly advocating for more democratic and equitable institutions and policies to help Nepal achieve the potential for green and just hills. The Federation of Community Forest Users Nepal (FECOFUN), for instance, represents 22,266 CFUGs as the largest network of grassroots stakeholders strongly advocates for local use rights. Similarly, the Himalayan Grassroots Women's Natural Resource Management Association of Nepal (HIMAWANTI) and Nepal Federation of Indigenous Nationalities (NEFIN) play crucial role in advocating for the rights of women and indigenous peoples concerning natural resources and forests.

Grey Hills (Early 2000 – Present)

On March 12, 2009, NASA released an image showing forest fires across the Himalayas of Nepal (National Aeronautics and Space Administration, 2021]). 2009 saw a record number of forest fires in Nepal (ICIMOD, 2024) – making the hills of Nepal almost grey with heavy smoke covering its lush green forests. Forest fires in Nepal have been the headline of many newspapers since the mid-2000s as the numbers and frequency of uncontrolled burns have been on the rise. In 2009, devastating forest fires ripped through 48 locations across Nepal, and in 2016 forest fires damaged more than 12,000 community forests in 50 districts. Between 2001 and 2019, more than 38,000 fire incidents occurred (ICIMOD, 2024). Once the green hills attracted global attention, Nepal's forests have now converted to grey hills with heavy smoke. Increased forest fires in Nepal have been linked to climate change. The Climate Change Policy 2019 blamed climate change for forest fires in Nepal (GoN, 2019) and predicted higher incidence of forest fires by 2065 (GoN, 2021) (Figure 9.2).

However, socio-economic and political processes are also contributing to forest fires in Nepal. First, the people-forest linkages have declined over the decades. The declining need for forest products (due to cement house construction and gas cooking stoves, among other factors) has significantly reduced community interest in community forests, leading to passive management of community forests (Ojha, 2023). Increased road access has brought liquefied petroleum gas (LPG) to villages, further lowering dependence on fuelwood. The changing livelihood strategies from subsistence farming to non-farm activities, especially remittance and wage labour, have decreased people's dependence on forest resources (Shahi et al., 2022).

Second, out-migration especially from the mid-hills has created a shortage of labour; the women, children, and elderly people staying back in the villages have

Figure 9.2 Forest fire in the Dadeldhura District of Nepal.

Photo credit: Jitendra Bajracharya_ICIMOD.

not been able to engage as actively in forest management (Poudyal et al., 2023). As a result, many community forests now have accumulated large and fire-prone fuel loads because forest cleaning and thinning activities have been stopped.

Justice and Equity Problems during the Grey Hills Phase

The entrenched inequity of the Green Hills of the 1990s has not abated in the Grey Hills era. Previously, the poor and socially marginalised communities actively and often voluntarily contributed to forest management activities. But most forest benefits went to elites because of their alliance with local leaders and bureaucrats. Access to forest benefits depended upon caste, income, and gender, with uneven distributive outcomes (Chaudhary et al., 2018). Socially and economically privileged groups often took the most forest benefits. Uneven distributive outcomes have led to frustration among the socio-economically disadvantaged groups leading to low sense of community and decline in active participation. In addition, with increased remittance income and low dependence on forests, the marginalised communities who had for years been guarding the forest and doing other labour-intensive works no longer feel obliged to actively participate in community forestry.

Conclusion

Community forestry has transformed the theories, policy, practice, and landscapes of Nepal. The decentralised approach not only contributed to an exceptional increase in forest cover but, in some cases, contributed to people's well-being and strengthened local leadership. The centralised approach during the Red Hills phase where the state and technocrats were controlling the forests resulted in degradation because the focus was more on profits, rather than supporting local people's livelihoods. Environmental injustices were rampant during the Red Hills era when the state controlled the forests and benefits flowed to the elites while burdens fell on the poor and the marginalised people. But under community forestry in the Green Hills access to forest resources was a great incentive for the local mountain communities to sustainably manage resources because the forest was the base of livelihoods and local economy.

And yet, environmental injustice has continued to plague community forestry in Nepal. During the Green Hills period, there have been problems of unequal participation of marginalised sections of society including women, ethnic communities, and Dalits. The failure of attaining just processes during the Green Hills era, especially the problem of restricted participation of marginalised groups in decision-making, laid a foundation for continued injustice in the current era of Grey Hills. To create environmental justice requires ending the marginalisation of certain sectors of society and creating fair and just processes and outcomes. This is critical for the future of community forestry in Nepal.

Acknowledgement

I would like to thank Director General Pema Gyamtsho, Srabani Roy, and Bandana Shakya of ICIMOD for allowing me to contribute to this important topic. I acknowledge the support from ICIMOD's core donors: the Governments of Afghanistan, Australia, Austria, Bangladesh, Bhutan, China, India, Myanmar, Nepal, Norway, Pakistan, Sweden, and Switzerland.

References

Agarwal, B. (2009). Gender and Forest Conservation: The Impact of Women's Participation in Community Forest Governance. *Ecological Economics*, 68(11): 2785–2799.

Baland, J., Bardhan, P., Das, S., & Mookherjee, D. (2009). *Forest Degradation in the Himalayas: Determinants and Policy Options*. Accessed on 5 January 2024 from https://www.isid.ac.in/~pu/conference/dec_09_conf/Papers/DilipMookherjee.pdf

Baral, S., Meilby, H., & Chhetri, B. B. K. (2019). The Contested Role of Management Plans in Improving Forest Conditions in Nepal's Community Forests. *International Forestry Review*, 21(1). https://doi.org/10.1505/146554819825863799

Basnyat, B., Treue, T., Pokharel, R. K., Kayastha, P. K., & Shrestha, G. K. (2023). Conservation by Corruption: The Hidden yet Regulated Economy in Nepal's Community Forest Timber Sector. *Forest Policy and Economics*, 149. https://doi.org/10.1016/j.forpol.2023.102917

Bhusal, P., Paudel, N. S., Adhikary, A., Karki, J., & Bhandari, K. (2018). Halting Forest Encroachment in Terai: What Role for Community Forestry? *Journal of Forest and Livelihood*, 16(1). https://forestaction.org/wp-content/uploads/2021/12/2-Bhusal_Halting.pdf

Blaikie, P. (1985). *The Political Economy of Soil Erosion in Developing Countries*. London: Longman Development Series.

Blaikie, P., Cameron, J., & Seddon, D. (2002). Understanding 20 Years of Change in West-Central Nepal: Continuity and Change in Lives and Ideas. *World Development*, 30, 1255–1270.

Chaudhary, S. (2017). *Paristhitiki Sewa: A Critical Analysis of Ecosystem Services Discourse in Nepal*. A PhD thesis submitted to Macquarie University, Australia.

Chaudhary, S., McGregor, A., Houston, D., & Chettri, N. (2018). Environmental Justice and Ecosystem Services: A Disaggregated Analysis of Community Access to Forest Benefits in Nepal. *Ecosystem Services*, 29(Part A). https://doi.org/10.1016/j.ecoser.2017.10.020

DFRS (2018). Forest Cover Maps of Local Levels (753) of Nepal. Department of Forest Research and Survey (DFRS), Kathmandu, Nepal.

Eckholm, E. (1984 Feb 21). *Seedings Dot Nepal's Once-Barren Slopes as Country Battles Forest Crisis*. The New York Times.

Eckholm, E. P. (1976). *Losing Ground: Environmental Stress and World Food Prospects*. New York: Norton.

FRTC (2022). *High Mountains and High Himal Forests of Nepal*. Forest Research and Training Centre (FRTC), Babarmahal, Kathmandu, Nepal. Accessed on 2 February 2024 from https://frtc.gov.np/uploads/files/HMHH_report%402022.pdf

Gautam, A., Shivakoti, G. P., & Webb, E. L. (2004). Forest Cover Change, Physiography, Local Economy, and Institutions in a Mountain Watershed in Nepal. *Environnemental Management*, 33(1), 48–61. https://doi.org/10.1007/s00267-003-0031-4

Gilmour, D. A., & Fisher, R. J. (1991). *Villagers, Forests and Foresters: The Philosophy, Process and Practice of Community Forestry in Nepal*. Kathmandu, Nepal: Sahayogi Press.

Giri, K. (2022). Is the Gender Agenda of Nepal's Community Forestry at Risk due to Commercial Transitions? *Forest Dynamics and Conservation: Science, Innovations and Policies*. https://doi.org/10.1007/978981-19-0071–6_6

GoN (2014). *Nepal National Biodiversity Strategy and Action Plan 2014–2020*. Kathmandu, Nepal: Ministry of Forests and Soil Conservation.

GoN (2019). *National Climate Change Policy 2076*. Government of Nepal. Accessed on 18 January 2024 from climatechangepolicyenglishpdf-9975-841-1660734867.pdf (mofe.gov.np)

GoN (2021). *Nepal's Long-term Strategy for Net-zero Emissions*. Government of Nepal. Accessed on 16 January 2024 from NepalLTLEDS.pdf (unfccc.int)

Hobley, M. (2012). *Persistence and Change: Review of 30 Years of Community Forestry in Nepal*. Kathmandu, Nepal: HURDEC.

ICIMOD (2024). *Putting Out Fires: Predicting and Curbing Forest Fire Damages in Nepal*. Accessed on 22 January 2024 from Putting out fires: Predicting and curbing forest fire damages in Nepal – ICIMOD

Ives, J. D., & Messerly, B. (1989). *The Himalayan Dilemma: Reconciling Development and Conservation*. London: Routledge.

Laudari, H. K., Sapkota, L. M., Maraseni, T., Subedi, P., Pariyar, S., Kaini, T. R., Lopchan, S. B., Weston, C., & Volkova, L. (2024). Community Forestry in a Changing Context: A Perspective from Nepal's Mid-Hill. *Land Use Policy*, 110. https://doi.org/10.1016/j.landusepol.2023.107018

Maharjan, M. R., Dhakal, S. K., & Thapa, K. (2009). Improving Benefits to the Poor from Community Forestry in the Churia Region of Nepal. *International Forestry Review*, 11, 254–267.

Malla, Y. B. (2000). Impact of Community Forestry Policy on Rural Livelihoods and Food Security in Nepal. *Unasylva*, 51(3), 37–45.

Malla, Y. B. (2001). Changing Policies and the Persistence of Patron-Client Relations in Nepal. Stakeholders' Response to Changes in Forest Policies. *Environmental History*, 6(2). https://doi.org/10.2307/3985088

Malla, Y. B., & Hobley, M. (2022). Community Forestry in the Changing Political and Social Context of Nepal. In *Routledge Handbook of Community Forestry*, edited by Janette Bulkan, John Palmer, Anne M. Larson, and Mary Hobley, 477–490. London: Routledge.

Myers, N. (1986). Environmental Repercussions of Deforestation in the Himalayas. *Journal of World Forest Resource Management*, 2, 63–72.

NASA (2021). *A Fierce Fire Season in Nepal.* National Aeronautics and Space Administration. Accessed on 5 February 2024 from A Fierce Fire Season in Nepal (nasa.gov)

Ojha, H. (2023). *Reviving Nepal's Community Forestry Success Story.* Nepali Times. Accessed on 5 January 2024 from Reviving Nepal's community forestry success story | Nepali Times

Ojha, H. R. (2006). Techno-bureaucratic Doxa and Challenges for Deliberative Governance: The Case of Community Forestry Policy and Practice in Nepal. *Policy and Society*, 25(2). https://doi.org/10.1016/S1449-4035(06)70077-7

Ojha, H. R., Banjade, M. R., Sunam, R. K., Bhattarai, B., Jana, S., Goutam, K. R., & Dhungana, S. (2014). Can Authority Change through Deliberative Politics? *Forest Policy and Economics*, 46, 1–9.

Oldekop, J. A., Sims, K. R. E., Whittingham, M., & Agrawal, A. (2018). An Upside to Globalization: International Outmigration Drives Reforestation in Nepal. *Global Environmental Change*, 52. https://doi.org/10.1016/j.gloenvcha.2018.06.004

Paudel, N. S., Ojha, H. R., Banjade, M. R., Karki, R., & Tamang, S. (2021). *Reframing Nepal' Community Forestry in the Changing Socioeconomic Context: A Collaborative Review of Policy and Practice.* Accessed on 5 February 2024 from Reframing-CF.pdf (enliftnepal.org)

Pokharel, R. K., & Tiwari, K. R. (2013). Good Governance Assessment in Nepal's Community Forestry. *Journal of Sustainable Forestry*, 32, 549–564.

Poudyal, B. H., Khatri, D. B., Paudel, D., Marquardt, K., & Khatri, S. (2023). Examining Forest Transition and Collective Action in Nepal's Community Forestry. *Land Use Policy*, 134. https://doi.org/10.1016/j.landusepol.2023.106872

RECOFTC (2022). *A Woman's Journey from Community Forestry to Politics*. RECOFTC Nepal News. Regional Community Forestry Training Centre for Asia and the Pacific. Accessed on 3 February 2024 from A woman's journey from community forestry to politics | RECOFTC

Shahi, N., Bhusal, P., Paudel, G., & Kimengsi, J. N. (2022). Forest-People Nexus in Changing Livelihoods Context: Evidence from Community Forests in Nepal. *Trees, Forests and People*, 8. https://doi.org/10.1016/j.tfp.2022.100223

Smith, A. C., Hurni, K., Fox, F., & Hoek, J. V. D. (2023). Community Forest Management Led to Rapid Local Forest Gain in Nepal: A 29 Year Mixed Methods Retrospective Case Study. *Land Use Policy*, 126. https://doi.org/10.1016/j.landusepol.2022.106526

Sunam, R. K., & McCarthy, J. F. (2010). Advancing Equity in Community Forestry: Recognition of the Poor Matters. *International Forestry Review*, 12, 370–382.

Thoms, C. A. (2008). Community Control of Resources and the Challenge of Improving Local Livelihoods: A Critical Examination of Community Forestry in Nepal. *Geoforum*, 39, 1452–1465.

Thwaites, R., Fisher, R., & Poudel, M. (Eds.). (2017). *Community Forestry in Nepal: Adapting to a Changing World* (1st ed.). Routledge. https://doi.org/10.4324/9781315445168
Yadav, B. D., Bigsby, H., & Macdonald, I. (2015). The Relative Distribution: An Alternative Approach to Evaluate the Impact of Community Level Forestry Organisations on Households. *Land Use Policy*, 42, 443–449. https://doi.org/10.1016/j.landusepol.2014.08.019

Part 3

Conflicts over River and Lowland Conservation

Part 3

Conflicts over River and Lowland Conservation

10 Protected Areas and Expendable Communities

Human-Wildlife Conflict Survivors and Unjust Compensation in the Koshi Tappu Wildlife Reserve

Dhirendra Nalbo

Introduction

Biodiversity conservationists applaud Nepal's accomplishment over the last 50 years. Nearly extinct and endangered animals have been preserved and protected to remarkable numbers: 750 rhinoceros, 355 Bengal tigers, over 170 elephants, and more than 490 Asiatic buffaloes. These and other endangered wildlife live in 20 different protected areas (PAs) in Nepal, which cover over 23% of the country's total landmass.

But there is another side to this success story. Human Wildlife Conflict (HWC) annually causes around 40 deaths, over 140 individuals sustain injuries, and hundreds of families lose their livelihoods as wildlife stray from the PAs, eat crops, and destroy buildings. Damage from animals impoverishes communities living near parks. Communities also face authoritarian approaches to PA governance under which the government turned indigenous land into wildlife parks. Consequently, many became landless and were forced to live without sufficient natural resources. Increased urbanization along with preexisting caste and class discrimination further weaken communities, limiting their economic opportunities.

It is not just human beings who suffer. The wild animals, too, die and get killed. Poachers remain active. Angry farmers retaliate against wildlife. In October 2021, a fully grown male elephant – notoriously painted as "man killer" known as "makuna" – was killed by villagers in a paddy field within a buffer zone area of the KTWR (here onwards Koshi Tappu).

The increasing frequency of HWC and its impacts on the local community calls for a radical reflection on the government's approach to PA governance. In this chapter, after analyzing Nepal's history of PAs, I will examine the government's monetary compensation program to address HWC from the perspective of local residents. I illustrate how the lengthy and arduous process for compensation inadvertently leads to what I call "invisible environmental injustice." Analyzing community resistance to environmental injustices, I shall discuss how such resistance has led to what Arago (2023) and Pinto's et al. (2023) call sustained "territorialized environmental justice."

DOI: 10.4324/9781003371175-13
This chapter has been made available under a CC-BY-NC-ND 4.0 license.

No place shows these dynamics better than Koshi Tappu. Nepal's first and biggest wetland. This PA sees some of the highest numbers of HWC across the country as well as deep environmental injustices. Curious tourists enjoy watching endangered species like Asiatic buffalos and a wide diversity of birds. In contrast, the communities who live near the PA experience landlessness, death, injury, trauma, and property and livelihood losses. I seek to bring to light the type of lived environmental injustice that local communities face and to help imagine an alternative redefining of justice moving forward in the frontlines of PAs. After all, the movement against environmental injustice begins both in Nepal and globally with grassroots and self-initiated efforts. We must learn from local communities (Cole & Foster, 2001; Ghimire, 2003; Murdock, 2021).

Methodology

The chapter focuses on understanding how the HWC survivors in the Koshi Tappu understand justice, particularly regarding compensation. Understanding their experiences and situating them within environmental justice is a central theme of the chapter.

This chapter is an outcome of three weeks long intensive qualitative research carried out in November 2017 and September 2022. I collected data through in-depth and semi-structured interviews, and participant observation in the key sites where the villagers experience conflict with wildlife. They include three municipalities – Kanchanrup, Barahkshetra, and Belaka – and one Gaupalika (Koshi Gaupalika). I interviewed fishermen, pastoralists, peasants, park officials, local government representatives, and PA committee officials.

The History of PAs in Nepal

When Nepal started its conservation effort, two phenomena were occurring at the same time: the absolute monarchy (1960s–1980s) and the Cold War (1950s–1980s).

When King Birendra assumed power in 1972 after his father King Mahendra died, Nepal's domestic politics was increasingly becoming volatile due to the rise of anti-monarchy forces. Internationally, the Cold War was gripping the world. For Western democracies, particularly the United States, it was an important period to engage with the Government of Nepal (GoN) to prevent it from getting too close to communist China (Robertson, 2018). For King Birendra to assuage domestic politics, gaining technical, financial, and political support from Western democracies by being part of the global environment movement was a strategic step (Croes, 2006).

It was also the time of the theory of Himalayan degradation (Eckholm, 1976). In it, Eckholm called for immediate intervention or face environmental destruction in the Himalayas. He implied that peasants' unsustainable and environmentally destructive farming techniques were the main culprit behind the region's environmental disasters. Western democracies and non-governmental organizations embraced the theory; the monarchy too responded to it by declaring six PAs and made *hariyo ban, Nepalko dhan* or "green forests are Nepal's wealth" its main development slogan.

For the PAs with sparse human settlements, like the Sagarmatha National Park (established in 1976) and Shey Foksundo (established in 1983), the authoritarian approach has arguably caused fewer sociopolitical and economic impacts. But for the PAs with dense human settlements, notably Chitwan National Park (CNP), Bardia National Park (established in 1976), and Koshi Tappu (established in 1976), the approach has had wider impacts on local communities. When the government established Koshi Tappu, it displaced 12,000 families (Shah, 1997).

The monarchy that introduced an authoritarian panchayat system of government "suitable to Nepali and Nepal's soil," disregarded the local communities' traditional knowledge about environmental conservation. Instead, it embraced the "wilderness" protection or fortress conservation model that the United States pursued and championed (Siurua, 2006). The international conservation partners too quietly accepted the government's authoritarian approach to establish PAs (McLean, 1999; Campbell, 2005, 2013).

By highlighting this historical background, I invite other researchers to consider why the long-term stewards of the environment – the local communities – were pushed aside and their "traditional ecological knowledge" ignored (Berkes, 1999).

The KTWR

In 1976, the GoN established the Koshi Tappu to protect the Asiatic Buffalo or *arna (bubalus bubalis)* (Shah, 1997). Covering 176 square kilometers and with an almost equal area demarcated as "buffer zone," this PA lies in Nepal's southeastern plain. The "Saptakoshi"– a mix of seven major rivers – flows through the PA, making it rich in biodiversity. But protecting this astounding diversity has come with an extremely high cost: human displacement, land dispossession, and frequent HWC.

Of the 12,000 families displaced from the PA, several hundred families who have been relocated to various parts of Morang and Sunsari districts have not received land ownership certificates. Thus far, the government has formed 21 Land Compensation Distribution Commission Committees. But all have failed to address the problem due to political, financial, and legal complexities. Consequently, the displaced population's future remains uncertain.

HWC adds another layer to environmental injustice in the Koshi Tappu. Of the total HWC incidents in Nepal's 20 PAs, at least two-thirds happen in the Koshi Tappu (Annual Reports, DNPWC 2076/77, 2077/78 and 2078/79). The people who endure deaths and injuries mostly come from economically impoverished, low (formally) educated and politically disempowered groups and face not only psychological pain but also an uncertain future due to cyclical poverty. Such structural inequalities prevent them from exercising their legal rights to seek compensation.

To address these issues, the government has introduced "people-centric" laws, policies, and regulations: they range from allocating budget for community development to promoting "ecotourism," and giving direct monetary compensation.

Human Wildlife Conflict-Compensation and Invisible Environmental Injustice in the Koshi Tappu

> Elephants are for those who have money to come and see. We protect them by allowing ourselves to be killed. When they come and attack us, we cannot hurt them. If we hurt them, we will go to jail. But when they attack, we will have to pay with our lives.
>
> (Personal communication with local resident, October 2018)

HWC repeatedly makes headlines in Nepal. Drawing from the International Union for Conservation of Nature (2023), I define HWC as direct and or indirect encounters between humans and wildlife resulting in injuries, deaths, or property and livelihood damage. Unfortunately, the government's effort to address HWC is causing more injustice than justice.

The quotation above is from Birendra whose father was killed in a 2014 elephant attack in their paddy field in the buffer zone area. "I was 26 when the incident took place," he recalled. "I had to learn everything, how to support my family, economically and otherwise" (Personal communication, November 2017).

In a 2017 incident, an elephant killed a woman named Yamuna while she was inside the park collecting grass for her goats. When asked about the incident, Nar, Yamuna's husband, stared at a distance and caressed the youngest of their three children. His neighbors filled in the silence describing how sorrowful life has been for Nar and his family since Yamuna's death. The government gave no compensation but Nar's family is not an exceptional case to not get compensation. In 2021, elephants killed two people while inside the park collecting fodder, and their families received no compensation either.

According to Nepal's conservation laws and policies, since the elephant killed Birendra's father outside the park, the family received compensation (Rs 500,000 or less than $5,000, after the repeated community protests, the government has increased the amount to $10,000). But since Yamuna and the other two individuals were killed inside the park; the deceased families were ineligible for compensation.

Based on their relatively comfortable economic condition, families like Birendra's rarely go into the park: they have enough land to produce food and can avoid the risk of going into the PA. But as daily wage laborers who struggle to feed their few goats, which provide critical nutrition and income, Nar's family must go inside the park since grass is not available outside. Consequently, families like theirs are more likely than Birendra's to get attacked by wildlife.

From a legal standpoint, the criteria for compensation may make sense. However, when the legal provisions are put in a socioeconomic context, they reveal what I call invisible environmental injustice. When laws and policies fail to respond to complex socioeconomic and political realities, communities experience invisible environmental injustice. For example, nearby residents must have a citizenship card to seek compensation for wildlife-caused losses. But many widows and single mothers often don't have citizenship and thus become ineligible for compensation.

In addition, those who are eligible to seek compensation often do not seek it because (a) they come from lower societal strata and want to avoid being humiliated by officials and (b) the compensation for livelihood losses is so small ($50–250) that the families decide it's not worth making the repeated visits to the government office necessary to attain compensation. On top of this, the monetary compensation for deaths and physical injuries does not address the trauma and other socioeconomic challenges that the survivors endure.

Certainly, the notion of justice does not end with compensation as an elected official shared, frustrated, "We cannot buy off grief or lives with 10 lakhs [rupees]. *Manchhe thulo ki janwar thulo?* [Who is more important, humans or the wild animals]" he asked. My interviewees believed that the government prioritized wildlife over them.

For families like Nar's, living next to the Koshi Tappu has turned into a death trap. They neither created this situation nor do they have an alternative way or means to escape it. Being born and raised in an area designated as a PA is their misfortune. Unfortunately, Koshi Tappu's legal framework makes it seem like the lives of ordinary people are disposable, mocking the very idea of justice and human rights.

Incidents involving property damage and livelihood loss also reinforce invisible environmental injustice. In 2022, across Nepal's 20 PAs, the DNPWC recorded 12,672 such incidents. Of these, 1,225 were in the Koshi Tappu. In 2021, there were a total of 8,455 such incidents, 886 of which occurred in Koshi Tappu. What is striking about these incidents is that those who lost property and crops often decide not to seek compensation and/or are not qualified because of legal and practical reasons.

The process of applying for compensation for such losses is long and arduous. The families must gather at least six to eight documents: land ownership and citizen certificates, evidence about the incident, recommendation letters from the ward office and buffer zone committee, and assessment letters from "experts" and officials. The communities find gathering these documents and applying for compensation discouraging, especially daily laborers and small farm holders who don't have political connections. "For five-six thousand rupees compensation, I cannot spend three to six months running around," a peasant whose rice seedlings were rampaged by an elephant said. "I'd rather work and earn more than ten thousand rupees for that time" (Personal communication, Sunsari, 2017).

Compensation related to HWC is complex. One way of delivering justice for the families' losses certainly is government-defined monetary compensation. But this one-time compensation, and the legal rigidity and arduous process of seeking compensation, leads to invisible environmental injustice. This type of injustice appears in the form of physical injuries, psychological traumas, and endless poverty: an outcome of the rigid legal framework perpetuated by what Park and Ruiz (2021) and Murdock (2021) call "environmental privilege."

The lengthy and complex process is not the only factor that discourages peasants from seeking compensation. Because of wildlife, farmers have stopped growing corn, wheat, and rice since these crops are costly to farm and attract wildlife such as *arna* and elephants. They have switched to vegetables such as pointed gourds, beans, tomatoes, chilies, and mustard greens that are not as costly to grow, are profitable, and do not attract wildlife. But the existing guidelines for compensation do not include these vegetables. They include only corn, wheat, and rice. In other words, the compensation laws and policies do not reflect local realities.

As GoN and conservation organizations celebrate conservation accomplishments in the Koshi Tappu, they rarely recognize the contributions and sacrifices that local communities make. Nepal's conservation model should be rethought so that both visible and invisible environmental injustices are addressed.

Grazing and Fishing as Forms of Resistance to Environmental Injustice

In the face of these environmental injustices, the communities are not sitting idle and waiting to be rescued. Instead, drawing from their intergenerational experience and knowledge, they continue making their presence seen and heard by defying the government's laws and policies. Their resistance is present in the form of grazing and fishing.

Various scholars have studied community resistance towards environmental injustice (Guha and Martinez-Alier, 1997; Guha, 2000; Holifield et al., 2010; Pellow, 2018; Arago, 2023; Pinto et al., 2023). They stress that resistance-led justice can take different forms. In this section, I conceive of resistance as a part of survival, finding meaning, and (re)claiming place or territory. I look at grazing and fishing as forms of resistance – a means, not an end goal, a community-initiated bottom-up approach without wider external institutional support – to combat invisible environmental injustice. In such contexts, the politics and process of resistance is a form of justice.

This conceptualization of resistance resonates with what Holmes (2014) calls "counter-territorialization" – an active assertion or claim over resources of a place – to counter the government's "territorialization." The political dynamics of grazing and fishing in the Koshi Tappu illustrate this in important ways.

Grazing: In the Kachanrup Municipality of Saptari district's Ward 1, during the mornings and evenings, cattle herds occupy the entire road that leads to Koshi Tappu. Keeping a few pairs of cattle or a dozen or more per family is a traditional practice for the Yadavs, one of the ethnic groups adjacent to the Koshi Tappu. My interviewees estimated that about 70% of Yadavs who live in the area raise cattle. By selling livestock and milk (at 80–90 rupees – or USD 60 cents per liter) they earn much-needed hard cash. Beyond economic importance, cattle are deeply intertwined with Yadav social status and cultural practice. For them, offering a glass of fresh milk or milk tea to their visitors is intimately linked with their identity of being Yadav.

Like other groups who keep cattle, Yadavs desperately need grazing space. Their parents and grandparents used the Koshi Tappu, which they called "*mrigban*" or "deer-forest," as pasture. When the government declared Koshi Tappu a PA, it provided no alternative grazing land.

The communities know that it is "illegal" to graze in the park, but out of necessity they still use it. For them, grazing is not simply continuing their traditional economic practice, but also an active way to assert their rights, voice, and social position. The outcome of this act of resistance is counter-territorialized environmental justice.

Fishing: "If the Yadav sees life in cattle, the Malah sees life in water," a fisherman from the ethnic Malah community shared with me. This community that lives in and nearby the park primarily depends on fishing for their sustenance. When the government created Koshi Tappu, it ignored the intimate relationship that the Malah has with the river and prohibited them from fishing.

"We used to form a group of four-five and sneak into the park," a 50-year-old fisherman noted. "We had to be careful and alert but when the officers would catch us, we had to give more than half of our catch or else face fine, jail or both" (Personal communication, 2018).

Only in 2009, 33 years after the creation of Koshi Taapu, did the government grant the Malah permission to fish inside the park. However, the fishing permission cards come with regulations. Fishing is allowed between 6 am to 6 pm. But in July-October it is prohibited because it is fish breeding season. "Fishing is like getting monthly salary for us," a fisherman said with reference to the regulation, "Can the warden live without receiving salary for three months?" He further bemoaned, "*Warden ko Koshi and warden ko lakadi* [The warden owns the river and its woods]" (Personal, communication, 2018).

Fishing for the Malah is more than an economic activity; it is linked to their history, culture and identity (Figure 10.1). Finding their life and rhythms with the Koshi River, which includes knowing flood patterns and fishery dynamics, distinguishes them from other communities. During my conversation with five elderly Malah men, they rarely expressed their frustration about the annual floods that the Koshi River brings. "The Koshi changes its course, so do we," a retired fisherman shared, "The Malah's life is in the Koshi River, and it ends with it" (Personal communication, August 2022).

Under the current legal framework, these communities are "illegal" settlers, because they do not have land ownership certificates. But the communities consider themselves the rightful residents of the area. "I am 77 years old and have been living here as long as I remember," a retired fisherman said. "My parents died in this house, and I raised my kids in this house, What about it makes me "illegal?" (Personal communication, August 2022).

By defying the government's laws and policies, the Malah fishermen endure baton blows, slaps, humiliation, fines, jail time, and confiscation of their catch and fishing tools. But through such active defiance, they maintain a form of their history, culture, and being: fishing. Seen in this light, like grazing for the Yadavs, fishing for Malah is a deliberate act of "counter-territorialization" (Holmes, 2014) through which they influence and defend a specific territory in the Koshi Tappu.

But counter-territorialized environmental justice cannot deliver real justice. Ultimately, the politically weak actors like the Malahs and Yadavs lack sufficient power to compel the government to reform its approach. Instead, because of their limitations, the community resistance ultimately remains "ineffectual," not "effectual" resistance (Malin et al., 2023).

Figure 10.1 A member of the Malah ethnic group fixing a fishing net.

Photo credit: Author.

Conclusion

Environmental injustice that occurs in relation to PA governance in Nepal requires urgent attention. Annually, dozens of people die due to HWC, hundreds sustain injuries, and thousands lose their livelihoods. As we celebrate Nepal's remarkable accomplishments in biodiversity conservation, we must remember these environmental injustices too.

For this accomplishment to continue, GoN and its partners must radically remodel its conservation approach that respond to the genuine needs and realities faced by local communities. We need to stop portraying an individual who acts as a human shield to protect livestock from a potential wildlife attack as a "foolish person" who needs to correct such a behavior. Rather we must seriously ponder what compels the person to make such a life-threatening decision and what can be done to reduce HWC in ways that promote just conservation.

Monetary compensation to address HWC is only an immediate step to what is a longer process of ending environmental injustice. Conflict is not simply between humans and wildlife; it is also between people. These conflicts have deep historical roots. Without addressing the structural factors that force humans and wildlife into conflict, "justice" focused on monetary compensation will only produce invisible environmental injustice.

Despite living under decades of hardship, the communities residing adjacent to PAs, such as those near Koshi Tappu, rarely protest against conservation. Nor have they demanded that PAs close. Instead, the communities want just equal treatment, like any other citizens, and recognition of their contribution to biodiversity conservation.

To address environmental injustices, the government and its conservation partners must pay careful attention to community grievances and must develop creative responses. Historic missteps can be corrected with targeted and contextualized interventions reflecting local realities. Doing so will require recognizing the community's contributions to conservation and involving them in a practical and tangible way based on their socioeconomic and cultural needs. As a concrete example, the government partnering with conservation organizations can find ways to provide free education for at least a generation of families living adjacent to the park who are not financially able to send their children to school. By taking such significant steps, we can prevent history from repeating, reduce the communities' pain and suffering, and recognize their sacrifice and contribution to Nepal's glorified biodiversity conservation accomplishments.

References

Arago, A. (ed.) (2023). *Transforming Spatial Data into Public Policies for Social Justice and Environmental Sustainability.* Cambridge Scholar Publishing.

Berkes, F. (1999). *Sacred Ecology: Traditional Ecological Knowledge and Resource Management.* Tylor and Francis.

Campbell, B. (2005). Nature's Discontents in Nepal. *Conservation and Society*, 3(22), 323–353.

Campbell, B. (2013). *Living between Juniper and Tree: Nature, Culture and the Power in the Himalayas.* Oxford University Press.

Cole, L. W., & Foster, S. R. (2001). *From the Ground Up: Environmental Racism and the Rise of the Environmental Justice Movement.* NYU Press.

Croes, K. D. (2006). Conserving the King: Inverting the Origin Story of the Annapurna Conservation Area Project of Nepal. *Himalaya* 26(1): 11–18.

Eckholm, E. (1976). *Losing ground: environmental stress and world food prospects.* New York: Norton.

Ghimire, S. (2003). Concept of Environmental Justice in Nepal: Environmentalism of Poor for Sustainable Livelihood. *Himalayan Journal of Sciences*, 1(1), 47–50.

Guha, R. (2000). *Environmentalism: A Global History.* New York: Longman.

Guha, R. & Martínez-Alier, J. (1997). *Varieties of Environmentalism: Essays North and South.* London: Earthscan.

Holifield, R., J., Chakraborty, & G. Walker, (eds.) (2017). *The Routledge handbook of environmental justice.* London: Routledge.

Holmes, G. (2014). Defining the Forest, Defending the Forest: Political Ecology, Territoriality, and Resistance to a Protected area in the Dominican Republic. *Geoforum* (53)1–10.

International Union for Conservation of Nature (2023). IUCN SSC Guidelines on Human-Wildlife Conflict and Coexistence. Available: https://portals.iucn.org/library/sites/library/files/documents/2023-009-En.pdf

Malin, S., Ciplet, D., & Harrison, J. (2023). "Sites of Resistance, Acceptance, and Quiescence Amid Environmental Injustice: An Introduction to the Special Issue on Sustainability Under Neoliberalism. *Environmental Justice* (16)1. 1–9.

Department of National Park and Wildlife Conservation (DNPWC), Kathmandu Nepal, "Wildlife Relief Distribution Guidelines."
Department of National Park and Wildlife Conservation (DNPWC), Kathmandu Nepal, Annual Reports. Available: https://dnpwc.gov.np/media/files/Final_Annual_Report_2078_79.pdf; https://dnpwc.gov.np/media/files/DNPWC_Annual_report-_2077.78.pdf; https://dnpwc.gov.np/media/files/DNPWC_Annual_Report-_077_Final.pdf; https://dnpwc.gov.np/media/files/DNPWC__Annual_Report-_074-75.pd
McLean, J. (1999). Conservation and the Impact of Relocation on the Tharus of Chitwan, Nepal. *Himalayan Research Bulletin*, 19(2), 38–44.
Murdock, E. (2021). "A history of environmental justice: foundations, narratives, and perspectives," in Coolsaet, B. (ed.) *Environmental Justice*. 6–18.
Park, L. & Ruiz, S. (2021) Racial Minorities in the United States, Race, Migration, and Reimagining, Environmental Justice in Coolsaet, B. (ed.) *Environmental Justice*. 225–234.
Pellow, D. (2017). *What is Critical Environmental Justice?* Cambridge, U.K.: Polity.
Pinto, A., et al. (2023). "Challenges Among Territorialization Among Ethnic, Peasant and Urban Waste Picker Peoples and Communities: The Case of Colombia," in Arago, A. (ed.) *Transforming Spatial Data into Public Policies for Social Justice and Environmental Sustainability*. Cambridge Scholar Publishing. 216–233.
Robertson, T. (2018). DDT and the Cold War Jungle: American Environmental and Social Engineering in the Rapti Valley of Nepal. *Journal of American History*, 104(4), 904–930.
Shah, J. P. (1997). *Koshi Tappu Wetlands: Nepal's Ramsar Site*. TUCN.
Siurua, H. (2006). Nature Above People: Rolston and "Fortress". *Conservation in the South Ethics & the Environment*, 11(1), 71–96.

11 The River People and the Parks

Political Ecology of Conservation and Indigenous Livelihoods in Nepal's Terai

Naya Sharma Paudel, Sudeep Jana Thing, and Rahul Karki

Introduction

The early establishment and expansion of PAs in Nepal was a combined project of global conservation actors and domestic ruling elites (Paudel, 2005). The conservationist state's imposition of the 'old protected area paradigm' (Stevens, 2014a) was largely driven by scientific knowledge, legal regimes, and the mobilization of the armed forces that excluded and marginalized pre-existing resident Indigenous Peoples and Local Communities (IPLC) who had close interactions with nature. It gradually undermined their knowledge, customary practices, and livelihoods. State interventions undermined IPLCs' legitimate rights and livelihoods, involving spatial, physical, economic, political, and cultural displacements and marginalization (Stevens, 2014a).

This chapter examines how the Nepali state, its allied actors, and associated discourses of biodiversity conservation, concomitant policies, and everyday practices have created conservation injustices for poor and marginalized Indigenous Peoples in and around PAs in the lowland Nepal. We examine the experiences of two distinct indigenous fishing communities: first, the Bote, Majhi, and Musahar, in and around the Chitwan National Park (CNP), south-central lowlands; and second, the Sonaha in the lower Karnali river delta, adjacent to Bardiya National Park (BNP), in the mid-western lowland Nepal. Our research questions are: How has the evolution and management of PAs affected livelihoods and cultures of Nepal's river people?, How are river people struggling to generate their livelihoods and sustain cultural practices?, How do conservation discourses and 'expert' knowledge help perpetuate injustice to Indigenous Peoples?, What of the three forms of environmental injustice – distributive, procedural, and recognition – that exist in relation to river people?

Through our research, we found that there has been some good progress since the 1990s in regard to access by IPLC to financial and capacity development as well as increased spaces of participation through local forest user groups and other institutions. However, fundamental gaps still exist in recognizing IPLC's worldviews, cultures, identity, and ways of life. Accordingly, we argue, despite changing conservation discourses and shifting away from protectionist paradigm toward participatory conservation, unequal power and knowledge between the state and

DOI: 10.4324/9781003371175-14
This chapter has been made available under a CC-BY-NC-ND 4.0 license.

IPLCs has helped maintain persistent injustice to these communities. Reframing conservation through the ideals of environmental justice can help redirect the policies and institutions toward a just conservation that recognizes IPLC's voices and ensures their legitimate rights, especially rights to manage and benefit from local resources.

Nature Conservation and Justice

We adopt a political ecology approach (Robbins, 2012) with an environmental justice perspective (Sze and London, 2008) to examine conservation injustices (Martin et al., 2015) around PA management. Nature conservation, especially through a PA approach, rearranges land and other natural resources and produces both winners and losers. As Martin et al. (2016) argue, conservation creates injustice because (1) PAs are often located in areas of high biological and cultural diversity with people whose knowledge and institutions may be marginalized and (2) the modern idea of PA emerged from the West and does not recognize Indigenous Peoples and their ways of conserving nature (Colchester, 2003) In the context of PAs, several dimensions of environmental justice – recognition, procedural, and distributional – are at stake.

Modern conservation was introduced in the 1960s in Nepal mainly through joint efforts of Western biologists and conservationists on one hand and Nepal's ruling elite on the other. Experts from international agencies such as UN Food and Agriculture Organization and London Zoological Society were able to convince Nepal's monarchy, who were avid hunters, on the imminent threats to the population of megafauna. As a result, PAs were established, expanded, and consolidated.

During the early 1960s, several studies raised alarms about declining wildlife and habitat destruction (Heinen and Kattel, 1992; Heinen and Shrestha, 2006). In the late 1960s, predictions of rhino extinctions by the late 1980s triggered a sense of urgency among political actors and media (Mishra, 2008). In 1969, six Royal Hunting Reserves in the lowland Terai and one in the mountains were established (Shah, 2002; Mishra, 2008) with exclusive rights to hunting for the royal family. Later with the promulgation of the National Park and Wildlife Conservation (NPWC) Act 1973, Chitwan National Park (CNP) was established. Crisis narratives of rhino and tiger population loss became the key rationale for both Nepal's ruling elite and the international conservation campaigners to establish and expand lowland PAs in Nepal (Thing, 2014). Within national-global conservation discourses, a network of actors and institutions – domestic ruling elites, forest bureaucracy, national and foreign experts, and international agencies – driven by scientific expertise, backed by external funding and technical assistance, provided an impetus to the creation of early PAs in Nepal (Thing, 2014).

Protected Areas and Environmental Injustice

This chapter is based on empirical data from two distinct indigenous river people living near two PAs – Bardiya and Chitwan, in Nepal. Insights are drawn from the authors' own engagement with these communities over the last three decades. Some of the reflections are also based on visits by the authors to these areas in early 2023.

River People and the Chitwan National Park

Chitwan Valley, located in central Nepal is endowed with fertile soil, dense forests, a huge river system, and wetlands. Indigenous people such as the Tharu, Bote, Musahar, Majhi, Chepang, Darai, and Kumal have been living there for generations. The Bote, Majhi, and Musahars live along the banks of the Narayani, Rapti, and Reu rivers. They used to fish, ferry people across the rivers, collect aquatic foods, and gather wild fruits and vegetables from the nearby rivers, wetlands, and forests. Because of the dense malarial forest, the Valley remained sparsely populated until the USAID-WHO project to eradicate malaria was introduced in 1960 (Muller-Boker, 2000). These indigenous groups all had limited natural immunity to malaria.

Historically, Chitwan was famous for wildlife hunting, especially Asian one-horned rhinos and Bengal tigers. Nepal's ruling class and their British allies often partook in the activity. However, with the growing poverty, environmental degradation, and especially the devastating floods in the nearby hill districts in 1954, the Nepal government partnered with WHO and USAID on a resettlement program, the 'Rapti Valley Development Plan', in 1956 (Robertson, 2018). Though the intent was to develop a model for integrated development, it created massive resettlement, heavy deforestation, and loss of biodiversity, especially wildlife (Ghimire, 1992) often with marginalization of local Indigenous People (Guneratne, 1998; Muller-Boker, 2000; Robertson, 2018).

By the late 1960s, half of the forests in the Valley were already cleared. As a result of the in-migration and concerns about poaching, conservationists convinced the Nepal government to promulgate the Wildlife Conservation Act of 1957. That same year, a Rhino Sanctuary was established in Chitwan, and later in 1961, a special force Rhino Patrol was set up to protect the mammals. In 1964, over 22,000 households were evicted or resettled outside the sanctuary. In 1967, a part of the Valley's forests was declared as wildlife sanctuary, and in 1973, it was finally turned into a national park (Paudel, 2005).

Since then, the Park authorities, supported by the Nepal Army, have cut off local access to resources by implementing a ban on collecting wild vegetables and fruits as well as grazing cattle inside the park, among others. The harassments and physical abuse against river people in the Valley continued in the 1990s and beyond (Amnesty International, 2021; Buzzfeed News, 2019). On 30 January 1993, CNP authorities raided several settlements during which boats and fishing nets were confiscated, torched, or destroyed. They also assaulted the river people. To this day locals still remember it as a dark day in their lives.

The NPWC Act 1973, defines 'fish' as a form of 'wildlife'. Fishing is equated with hunting or poaching, and therefore is prohibited. The PA authorities usually blame the region's river people, claiming that their fishing is illegal, and thus threatens the biodiversity. However, the river people have their own story, as a Majhi man from Nawalparasi explained to us:

> Our ancestors lived here for centuries with the wild animals. The Park came only during our time; a few decades ago. The government also brought *vikas* [development] with the park. The cement factory, paper factory, all pour their waste into the rivers. The big dam in Indo-Nepal border blocks fish movement. Who is damaging the nature? We indigenous people or the *Sarkar* [government]?

The river people also argue that their movement along the rivers for fishing actually deters poachers. A Bote activist argued:

> While we used to roam frequently along the Narayani for fishing, no one dared to come with an ill intent. Now the area is rather quiet. Poachers use such an opportunity to get into the jungle, watch the wildlife movement and plan their illegal activities.

These arguments have been supported by anecdotal studies. However, there is a severe lack of credible research on many of these contested issues to help inform the dialogue.

With no land entitlement and private property, the restriction on access to fishing, ferrying, and collection of wild food has severely undermined the livelihoods of river people. Many families, especially their children are increasingly renouncing traditional occupations and are opting for wage labor or outmigration. As it is, their concerns rarely find support in any political discourse, election agenda, or civic movements.

In 1996, the Government introduced participatory conservation in the form of the Buffer Zone Management Program (BZMP). It had two major objectives: (i) delineate land and resources adjacent to PA and manage them sustainably to reduce the local people's resource extraction pressure from PA and (ii) invest 30%–50% of the PA revenue to the socio-economic development of neighboring communities which will gradually lead to decreased reliance on forest resource. However, contrary to its stated intent, the BZMP has been unable to benefit river people (Paudel et al., 2007; Rai et al., 2023). While forest protection and the increased availability of forest biomass may have benefitted smallholder farmers, the fishermen continued to face restrictions in fishing and collection of wild food items. Also, while local elites have benefitted from mobilization of buffer zone funds, river people are often excluded or marginalized from buffer zone institutions, decision-making and benefit sharing.

Instead, river people face increased scrutiny from local buffer zone leaders, questioning their traditional livelihood practices and increased PA-local elite alliance that further delegitimizes their culture and traditional way of life.

Despite being closely monitored by armed security personnel, river communities have engaged in diverse covert and overt forms of resistance. They do this in several ways: by collectively occupying public land, sheltering on the river islands in the buffer zone, and secretly trespassing park boundaries at night. They also trick guards during their patrolling, bribe officials and guards with fish catch or wild food items and provide them with free labor as well as guide their way into the forests.

In recent decades, river people have adopted other strategies too: expanding market and civil society activities; and increasing adoption of confrontational open political actions such as protest rallies, mass meetings, press releases, and lobbying with political parties (Paudel, 2005; Thing, 2020). In 1992, they formed an organization called the Majhi, Musahar, Bote Welfare and Service Committee, organized several internal and stakeholder meetings, made several delegations to the park headquarters, staged protests and rallies, and engaged in negotiation with the authorities. In 1999 they encircled the park headquarters demanding reinstatement of fishing rights and a public hearing to voice their concerns. Many of them now have a fishing license for five months (September-January) a year. Their struggle was partly supported by NGOs (Jana, 2007).

The Sonaha of BNP

The marginalized ethnic minority known as the Sonaha consider themselves indigenous to the lower plains of Bardiya in the western lowlands. Despite their unique ethnic identity and culture, they are still struggling to gain legal recognition. According to the 2011 Census, there are 579 people with Sonaha as their mother tongue (CBS, 2011), although this figure is likely an undercount (Thing, 2014). The Sonaha reside in several rural settlements in the vicinity of the Karnali River and its branch Geruwa, in the lower Karnali River floodplains. This is situated in the western section of the BNP and its buffer zone.

The Sonaha traditionally relied on artisanal fishing and gold panning on the riverbanks, as well as access to forest resources for their livelihood. Their ancestors also engaged in ferrying across the rivers. Despite the availability of fertile land in the region, the ancestors preferred a semi-mobile way of life to agrarian. "*Swan Machhi Kheti*" (gold and fish as cultivation) and "*Tipariya ma bashahi*" (temporary shifts on the river islands) define their customary livelihoods and cultural identities. Across the riverbanks, Kafthans – a complex system of managing gold panning through a sacred common regulated by clan elders through the possession of animistic shrines – were typical in the past (Thing, 2019). However, the seminomadic ways of life have gradually declined in recent years in favor of subsistence farming, seasonal outmigration, and daily wages.

Bardiya National Park (BNP), the largest lowland PA of Nepal, has a history similar to the Chitwan National Park (CNP). The creation of BNP, a former hunting reserve of the Nepal's Royals, was largely shaped by discourse of protecting tigers as endangered megafauna in the Bardiya lowland forests. It was augmented by the nexus of powerful actors and institutions including the Royal family, the forest bureaucracy, and foreign wildlife experts (Thing, 2014). In 1996, areas and settlements adjacent to the core area of the BNP were declared as a buffer zone. In the early days of its establishment, when forest guards were deployed, there was still some level of tolerance of the Sonaha in the river but that changed with the gradual deployment of military. Like the CNP, BNP was also a top-down imposition on the Sonaha lives and territory (Thing, 2014; Thing et al., 2017). The awareness about the creation of BNP and its policies and restrictions came through their encounters with army and ranger patrols, interception of their canoes and makeshift shelters on the river islands, confiscation of their fish catch, and fishing and gold panning tools.

The BNP management and policies have undermined the Sonaha control and access to their riverine territories, including customary livelihoods and cultural practices. A young male member of Sonaha community recalls:

> Sonaha had freedom over the rivers.... When it became *Arakshya* (wildlife reserve) everything stopped: hunting, fishing, and mobility. The army told us, 'Do not enter the forest'. If they found us there, they would arrest and take us away. Arakshya is for wild animals, we cannot go there.

In a similar story, an adult female member of the Sonaha shared:

> Gradually, we began to see soldiers guarding the forests across the river. We were told we could not catch fish in the river and enter the forest. If we were caught fishing, we would be fined and punished.

The Sonaha too have been resisting the BNP authorities and policies through silent, passive resistance as well as organized, collective resistance (Thing, 2014, 2020). After a persistent campaign and collective actions, the Sonaha managed to negotiate fishing and gold panning permits from BNP administration but it was short-lived. It was revoked after three Sonaha youths were found with a rhino horn. At the discretion of the BNP officials, on rare occasions, some Sonaha have received concessions to access certain portions of the rivers for fishing and gold panning; however, restrictions on access to rivers remain intact. Advocacy and campaign under the banner of Nepal Sonaha Association have fizzled out in recent years, but through the Sonaha Development Society, a Sonaha NGO, their struggle to be legally recognized as an ethnic group continues.

Continuation of Injustice: The Limitations of Participatory Conservation

The introduction of a participatory conservation approach in the mid-1990s that originally aimed at correcting the historical injustice and supporting welfare of the PAs-affected IPLCs, also largely failed to meet its mission. Participatory conservation's conservation biases, techno-bureaucratic approach, and non-recognition of indigenous worldviews and ontologies resulted in the continuation of fortress conservation. While the PA authorities were able to garner local support in exchange for this program, river people and other disadvantaged groups were often excluded or marginalized in buffer zone institutions. They have no influence when it comes to deciding on program priorities and resource allocations (Paudel et al., 2006; Paudel et al., 2007; Shahi et al., 2023). No representative from their community has yet been able to get to the Buffer Zone Management Council (BMC), the apex governance body. There is rarely any member from these communities in local decision-making institutions. Only two members from each local user group can vote for BMCs. Moreover, the preference of candidates is usually influenced by local political party leadership, often the largely dominant ethnic groups and men. As a result, the Buffer Zone Management Program became one additional instrument to further legitimize and perpetuate injustices to these river people (Paudel et al., 2006; Thing et al., 2017).

Despite some progress in participatory conservation approaches – for instance, local participation in conservation and development initiatives and institutions – decisions have mostly tipped the power toward the PA authorities and local elites and further marginalized river people (Ghimire, 2009; Paudel et al., 2012; Neupane and Majhi, 2016). Participatory conservation in Nepal has been criticized for being top-down and expanding state authority upon civic life (Heinen and Shrestha, 2006), for re-entrenchment of conventional authority (Ojha et al., 2014), for breeding conditions for elite capture (Adhikari and Lovett, 2006; Paudel et al., 2006), for aggravating conflicts at the local level (Paudel et al., 2006; Paudel et al., 2010), and for failing to address Indigenous Peoples' engagement with their environment (Campbell, 2005b). In fact, there is a continuation of the 'old paradigm' (Stevens, 2014a; Rai et al., 2023) even within the participatory regime.

Another limitation of BZMPs is that the policies and programs continue to embrace the dominant views on nature-culture division and ignore the worldviews, customary ways of living, and ancestral riverine territories of river people (Thing et al., 2017; Rai et al., 2023). The landscape is strictly divided into core zone, buffer zone, community forestry, wetlands, or other management regimes. Experts and bureaucrats lead the participatory approach with promotional activities and procedural compliances that contribute to depoliticizing resource access. Even the Buffer Zone leaders approve of the discourse of conventional conservation and use their authority to impose sanctions against fishing, gold panning, and collection of wild food items, among others. Alternative livelihood projects prioritize teashop, cycle repair, driving, plumbing, wiring, and tailoring, activities that do not fit culturally into the lives of river people.

In the meantime, a growing middle class and its alignment with modern environmentalism have attempted to further delegitimized the river people's ways of life. Nepal's urban elites know little about the violent history of the CNP and even less about the sustainable livelihoods of river people. Instead, environmentalist discourse celebrates rich biodiversity, and wildlife, especially the megafauna like rhinos and tigers. Nepal has made doubling tigers a national goal (which it has achieved by nearly tripping its big cat population to 355) but few give much thought to how PAs have undermined the rights and lives of river people. Consequently, park-people conflict continues and injustices are entrenched.

The river people often assert that through their cultural and ecological knowledge and wisdom, they can contribute as stewards of river conservation (Mustafa et al. 2021). In addition, they now have begun to articulate their struggle with wider national and international discourses of human rights and rights of Indigenous People. Some of them are now part of the National Federation of Indigenous Nationalities of Nepal and use the multilateral agreements such as ILO-169 and UN Declaration on Rights of Indigenous Peoples to make their case. Yet, the Sonaha continue to struggle for legal recognition of their ethnic identity and livelihood access claims.

Conclusion

PAs in Nepal's lowland Terai region were established as part of a global wave of PA expansion that aimed at protecting nature by excluding people. The unequal power dynamics between the ruling elites and experts on one hand and the marginalized IPLCs on the other made it possible to impose 'coercive conservation' (Peluso, 1993) and spatial control that significantly undermined the lives of indigenous river people. The river people are not simply passive victims; instead, they are active agents in material and discursive struggle against state power, experts, and conservation NGOs. While the degree of local consultations and participation in Buffer Zone Management Planning (BZMP) have been greater in recent years, planning processes are largely driven by PA officials and technical expertise. Participatory conservation fails to embrace indigenous knowledge and worldviews and falls short of 'a new conservation paradigm' (Stevens, 2014b).

In conclusion, if we are to address conservation injustice in and around PAs, we must recognize the historical disservice against river people and respect their alternative worldviews, their ways of perceiving and valuing nature, and their institutions and cultural practices associated with resource management and use. This means moving beyond distributive models of conservation costs and a participatory and inclusive process. This means changing power dynamics through institutional and structural reforms (Martin et al., 2016); assuming responsibility by those who are in position of power and reshaping relations (Young, 2010); and facilitating intercultural dialogue (Escobar, 1998; Dryzek, 2000) for a just conservation that works for both nature and people.

References

Adhikari B., Lovett J. C. (2006). Transaction costs and community-based natural resource management in Nepal. Journal of Environmental Management 78(1): 5–15.

Amnesty International. (2021). *Violations in the Name of Conservation*. Amnesty International and Community Self Reliance. London.

Buzzfeed News. (2019). WWF's Secret War. https://www.buzzfeednews.com/article/tomwarren/wwf-world-wide-fund-nature-parks-torture-death. Accessed on 18 May 2023.

Campbell, B. (2005b). Nature's Discontents in Nepal. *Conservation and Society*, 3(2), 323–353.

CBS. (2011). *National Population and Housing Census 2011*. Kathmandu: Central Bureau of Statistics (CBS).

Colchester, M. (2003). *Salvaging Nature: Indigenous Peoples, Protected Areas and Biodiversity Conservation*. Montevideo: World Rainforest Movement and Forest People's Programme.

Dryzek, J. S. (2000). *Deliberative Democracy and beyond: Liberals, Critics, Contestations*. Oxford: Oxford University Press.

Escobar, A. (1998). Whose Knowledge, Whose Nature? Biodiversity, Conservation, and the Political Ecology of Social Movements. *Journal of Political Ecology*, 5, 53–82.

Ghimire, K. B. (1992). *Forest or Farm? The Politics of Poverty and Land Hunger in Nepal*. New Delhi: Oxford University Press.

Ghimire, S. (2009). Samrakchyan ma Antarbirodh, Pratirodh tatha Bikalpa [Contradictions, Resistance and Alternatives in Conservation]. In H. Dhungana, S. Ghimire, & J. Adhikari (Eds.), *Jaiwik Biwidhata Ra Jiwika* [Biodiversity and Livelihoods] (pp. 119–142). Kathmandu: Martin Chautari.

Guneratne, A. (1998). Modernization, the State, and the Construction of a Tharu Identity in Nepal. *The Journal of Asian Studies*, 57(3), 749–773.

Heinen, J. T., & Kattel, B. (1992). Parks, People, and Conservation: A Review of Management Issues in Nepal's Protected Areas. *Population and Environment: A Journal of Interdisciplinary Studies*, 14(1), 49–84.

Heinen, J. T., & Shrestha, S. K. (2006). Evolving Policies for Conservation: An Historical Profile of the Protected Area System of Nepal. *Journal of Environmental Planning and Management*, 49(1), 41–58.

Jana, S. (2007). Working Towards Environmental Justice: An Indigenous Fishing Minority's Movement in Chitwan National Park, *Nepal*. Lalitpur. International Centre for Integrated Mountain Development (ICIMOD).

Martin, A., Akol, A., & Gross-Camp, N. (2015). Towards an Explicit Justice Framing of the Social Impacts of Conservation. *Conservation and Society*, 13(2), 166–178.

Martin, A., Coolsaet, B., Corbera, E., Dawson, N., Fraser, J., Lehmann, I., & Rodriguez, I. (2016). Justice and Conservation: The Need to Incorporate Recognition. *Biological Conservation*, 197, 254–261.

Mishra, H. (2008). *The Soul of the Rhino: A Nepali Adventure with Kings and Elephant Drivers, Billionaires and Bureaucrats, Shamans and Scientists, and the Indian Rhinoceros*. Guildford, CT: The Lyons Press.

Muller-Boker, U. (2000). State Intervention in Chitwan: On the Historical Development of a Region in Southern Nepal. *Studies in Nepali History and Society*, 5(2), 173–200.

Mustafa, D., Nayupane, G. C., Shrestha, K., & Buzinde, C. N. (2021). Scalar Politics of Indigenous Waterscapes in Navajo Nation and Nepal: Conflicts, Conservation and Development. *Environment and Planning: E Nature and Space 0(0)*, 1–23. Available at: https://journals.sagepub.com/doi/pdf/10.1177/25148486211007853. Accessed on: 11 Aug, 2024.

Neupane, C., & Majhi, C. (2016). *Samrachit Chetra Ra Dwanda (Protected Areas and Conflict)*. Nawalparasi and Chitwan: Majhi Musahar Bote Kalyan Sewa Samiti and Pravat Kiran Sewa Samaj.

Ojha, H. R., Banjade, M. R., Sunam, R. K., Bhattarai, B., Jana, S., Goutam, K. R., & Dhungana, S. (2014). Can Authority Change through Deliberative Politics? Lessons from the Four Decades of Participatory Forest Policy Reform in Nepal. *Forest Policy and Economics*, 46, 1–9.

Paudel, N. S. (2005). Conservation and Livelihoods: Exploration of Local Responses to Conservation Interventions in Royal Chitwan National Park, Nepal (Unpublished doctoral dissertation). The University of Reading, Reading, United Kingdom.

Paudel, N. S., Ghimire, S. and Ojha, H. R. (2006). Human rights: A guiding principle or a challenge for transforming conservation? *Policy Matters*, 15, 299–310.

Paudel, N. S. Sharma U. R. and Budhathoki, P. (2007). Buffer zones: New Frontiers for Participatory Conservation? *Journal of Forest and Livelihood,* 6 (2), 44–53.

Paudel, N. S., Jana, S., & Rai, J. K. (2010). Protected Areas and Rights Movements: The Inadequacies of Nepal's Participatory Conservation. ForestAction, Nepal, Discussion Paper Series, 10:3.

Paudel, N. S., Jana, S., & Rai, J. K. (2012). Contested Law: Slow Response to Demands for Reformulating Protected Area Legal Framework in Nepal. *Journal of Forest and Livelihood*, 10(1), 88–100. http://doi.org/10.3126/jfl.v10i1.8603

Peluso, N. L. (1993). Coercing Conservation? The Politics of State Resource Control. *Global Environmental Change*, 3(2), 199–217.

Rai, I. M., Melles, G., & Gautam, S. (2023). Community Development for Bote in Chitwan National Park, Nepal: A Political Ecology of Development Logic of Erasure. *Sustainability*, 15(2834), 1–19. https://doi.org/10.3390/su15032834

Robbins, P. (2012). *Political Ecology: A Critical Introduction*. Chichester: J. Wiley and Sons.

Robertson, T. B. (2018). DDT and the Cold War Jungle: American Environmental and Social Engineering in the Rapti Valley of Nepal. *Journal of American History*, 104(4), 904–930.

Shah, G. B. B. (2002). Search of a New Value System: Environmental Challenges in Nepal. In V. Menon & M. Sakamoto (Eds.), *Heaven, Earth, and I: The Ethics of Nature Conservation in Asia* (pp. 113–125). New Delhi: Penguin in association with International Fund for Animal Welfare.

Shahi, K., Khanal, G., Jha, R.R., Bhusal, P., Silwal, T. (2023). What drives local communities' attitudes towards the protected areas? Insights from Bardia National Park. *Conservation Science and Practice*, 5 (1–11).

Stevens, S. (2014a). Indigenous Peoples, Biocultural Diversity, and Protected Areas. In S. Stevens (Ed.), *Indigenous Peoples, National Parks, and PAs: A New Paradigm Linking Conservation, Culture, and Rights* (pp. 15–46). Tucson Basin: The University of Arizona Press.

Stevens, S. (2014b). New Protected Area Paradigm. In S. Stevens (Ed.), *Indigenous Peoples, National Parks, and PAs: A New Paradigm Linking Conservation, Culture, and Rights* (pp. 47–83). Tucson Basin: The University of Arizona Press.

Sze, J., & London, J. K. (2008). Environmental Justice at the Crossroads. *Sociology Compass*, 2(4), 1331–1354.

Thing, S. J. (2014). *The Polemics and Discourse of Conservation in Nepal: A Case Study of Sonaha Indigenous Minorities and Bardia National Park* [Doctoral dissertation, Curtin University]. http://hdl.handle.net/20.500.11937/912

Thing, S. J. (2019). Politics of Conservation, Moral Ecology and Resistance by the Sonaha Indigenous Minorities of Nepal. In C. Griffan, R. Jones, & I. Robertson (Eds.), *Moral Ecologies – Histories of Conservation, Dispossession and Resistance* (pp. 37–58). Cham: Palgrave Macmillan.

Thing, S. J. (2020). Riverscape as Biocultural Heritage: A Local Indigenous Social Movement Contesting a National Park in Nepal. In A. Mozaffari & T. Jones (Eds.), *Heritage Movements in Asia. Cultural Heritage Activism, Politics, and Identity* (pp. 123–146). New York and Oxford: Berghahn.

Thing, S. J., Jones, R., & Birdsall Jones, C. (2017). The Politics of Conservation: Sonaha, Riverscape in the Bardia National Park and Buffer Zone, Nepal. *Conservation & Society*, 15(3), 292–303.

Young, I. M. (2010). *Responsibility for Justice*. Oxford: Oxford University Press.

Part 4

Infrastructure and Indigenous Peoples

12 Disaster is Social

Uneven Effect and Recovery from the 2015 Nepal Earthquake

Mukta S. Tamang

Introduction

On 25 April 2015, an earthquake of 7.8 magnitude shook Nepal, with its epicenter northwest of Kathmandu. This was followed by a second earthquake with 7.3 magnitude on 12 May 2015. The earthquakes hit 31 districts in Nepal's central mountains, out of which 14 were categorized as highly affected. The earthquakes killed and injured people and destroyed physical infrastructure – houses, animal sheds, cultural monuments, community buildings, irrigation canals, and land. According to the Nepal Reconstruction Authority (NRA), 8,970 lives were lost and 22,300 people were seriously injured. In total, the earthquakes destroyed or damaged approximately 800,000 houses. A government geological survey identified 136 settlements needing complete relocation. (NRA 2018). The NRA estimated 750 cultural heritage sites and 1,200 monasteries were damaged, as well as 8,680 school buildings and 1,197 health institutions. The Post Disaster Needs Assessment (PDNA) by the government estimated that the disaster pushed an additional 5.4 million people below the poverty line (NPC 2015).

Located at the remote northern corner of Sindhupalchowk district, Baruwa village suffered greatly from the earthquakes. Almost all houses collapsed and 121 people lost their lives. Earthquake-induced landslides ripped scars on the village's rugged slopes. Maize crops were ruined. Afterward, survivors lived in tents as they organized funerals for dead loved ones. The villagers shifted to temporary shelters in the valley below and struggled to rebuild houses.

A poverty-stricken 35-year-old Tamang mother, who lost her husband in the earthquake, felt abandoned. Her husband had worked as wage laborer to supplement the limited yields they could squeeze from high-altitude terraced fields. Still living in the temporary shelter, she remarked,

> I almost gave up hope of building a new house as I do not have the means to organize labor and materials for construction. The government support is inadequate and I have no money to add. I worry about my two children.
> (Personal communication 2019)

DOI: 10.4324/9781003371175-16
This chapter has been made available under a CC-BY-NC-ND 4.0 license.

Even four years after the monstrous earthquake devastated communities in the central hills of Nepal, people – especially those facing pre-existing social discrimination – continued to struggle for recovery.

In the 2015 earthquake, Nepal's elderly, children, single women, individuals with disabilities, and other people in vulnerable circumstances faced the highest challenges. Given Nepal's history of caste, ethnic, and class-based discrimination, historically oppressed communities such as Indigenous Peoples, mostly Tamangs, Dalits, and other minorities encountered multiple obstacles. The recovery process has been uneven so far, some have bounced back better while others have struggled, with some finding themselves much worse off than before.

This uneven recovery came on top of the unequal results of the original disaster. The 2015 earthquakes disproportionately hit historically marginalized groups. Natural calamity is not neutral. Nepal's earthquakes undercut the poor and historically marginalized people more than others. Structural inequalities shaped the disaster's direct impacts as well as the recovery; these pre-existing inequalities often weakened the resilience capacity of communities.

Disasters in Nepal are often seen as fateful events (the term in Nepali is '*daibi-prakop*' or 'divine wrath') hitting all people equally, irrespective of class, ethnicity, caste, gender, and geographic location. While these are called these "natural" calamities, in reality, political-historical conditions and state action profoundly shape their outcomes. As Nepal has a past legacy of exclusion and discrimination which produced systemic inequalities and vulnerabilities based on class, caste, ethnicity, and gender (Bennett and Parajuli 2012; Höfer 2004; Lawoti 2008; Riaz and Basu 2007), the unequal impact and uneven recovery from the 2015 earthquakes show disaster injustice is yet to be addressed in meaningful ways. This paper demonstrates the unequal effect of the disaster, the uneven recovery, and the disproportionately distributed household capacity required for bouncing back.

The concept of disaster justice has been taken up by disaster researchers as a useful framework for understanding disaster risk preparedness, response, and recovery (Adhikari et al. 2023; Bhattarai 2018; Douglass and Miller 2018; Eda 2015; Shrestha et al. 2019; Verchick 2012). Disaster justice stresses that the impacts of disaster always have political causes and effects. It is human interventions that cause natural and environmental hazards that amplify their socially unequal impacts. Disaster justice highlights that marginalized communities (1) are most vulnerable to adverse effects and (2) face significant challenges to recovery (Cutter 2006; Flacke et al. 2022; Reid 2013; Robertson 2017). Disaster justice also sees the state as having mandatory responsibility to protect the vulnerable through inclusive policies and programs. Disaster justice is linked with the "rights of citizens to make claims through such established systems of governance to ensure fairness in the distribution of resources and services to prevent, respond to, and recover from disasters" (Douglass and Miller 2018: 275). In line with the Rawlsian principle of justice (Rawls 1971), disaster justice is a moral and legal claim on the state and system of governance for the protection of its vulnerable (Bankoff 2018).

The data for this paper comes from research conducted in 2019 on community resilience capacity in the context of the 2015 earthquakes through the Central Department of Anthropology at Tribhuvan University (Tamang 2020). The study combined quantitative and ethnographic/qualitative methods. A survey (Tamang 2020) was conducted with 3,300 households sampled in the 14 most earthquake-affected districts: Bhaktapur, Dhading, Dolakha, Gorkha, Kathmandu, Kavrepalanchowk, Lalitpur, Makawanpur, Nuwakot, Okhaldhunga, Ramechhap, Rasuwa, Sindhuli, and Sindhupalchowk. The caste and ethnic distribution of the sampled households roughly reflected the area's social diversity. The data was disaggregated to produce a comparative picture by gender, caste, ethnicity, and by income levels of the impact of and recovery from disaster.

The Past and Shaping of Vulnerability

The 2015 earthquake mostly hit the central mountains of Nepal. More than half of the population belongs to Indigenous communities followed by hill Brahman Chhetri and Dalits. The Indigenous Tamangs are the most numerous group. The population of Dalits in the region is about 6% (CBS 2014).

Before the formation of Nepal through the forceful conquest by the king of Gorkha in the middle of the 18th century, this region enjoyed autonomous polities. The Shah-Rana dynasty then ruled the country for more than two and half centuries. The state focused on the extraction of revenue from subjects. State "law and order" served to maintain domination and disregarded the welfare of the people and the development of the country. The ruling elites established themselves as landlords by dispossessing Indigenous peasants and inculcated a feudal ethos of loyalty, favoritism, and nepotism. The caste system segregated people into caste hierarchies based on the principle of purity and impurity.

Indigenous Tamang and Dalits suffered from the state policies in profound ways. The Muluki Ain, civil code established in the 1850s categorized all Indigenous groups as lower caste alcohol drinkers or *matwali*, and some, including the Tamang, as enslavable. Their fertile lands were taken away for *birta* and *jagir* land grants to Gorkhali leaders and soldiers. In addition, they were forced to give free labor to produce gunpowder, paper, and other materials required for the state (Holmberg 1999). The new bureaucracy displaced Tamang self-governance systems. Tamang were not allowed to hold any position of influence at the regional or national level (Tamang 1992). Further, the rulers established Hinduism as the state religion with Khas as the dominant Nepali language and gave official sanction to hill Brahmin/Chhetri cultural traditions (Tamang, 2009).

This history of dispossession and deprivation is reflected in inequitable human development status. Nepal's 1998 Human Development Report showed that the districts of Sindhupalchowk, Dhading, Rasuwa, and Sindhuli – where the majority of Tamang live – had a human development index comparable to that of the Karnali region (NSAC 1998) – the most deprived region in the country. Fifty-nine percent of Tamang lived in absolute poverty; the national average was 36%. Tamangs had a low literacy rate and a high infant mortality rate. Tamang's representation in political decision-making bodies was almost non-existent. These conditions are largely unchanged today.

Dalits were categorized by the Muluki Ain 1854 as *pani nachalnya* or "water unacceptable" low caste. Treated as "untouchable," they were subjected to strict rules preventing them from interacting with the upper castes. Although legal equality of all citizens was formally announced in the 1960s, the traditional caste place of Dalits at the bottom of the social hierarchy persisted (Bhattachan et al. 2009). The majority of Nepal's Dalits are landless and their livelihood is based on artisanal work and agricultural wage labor. About two-thirds of Dalits live in absolute poverty (NSAC 1998). Dalits have had limited access to education and Dalit children often faced discrimination from teachers and peers in schools. Dalits sometimes encounter violence. The 1990 peoples' movement brought multi-party politics and basic freedoms but the entrenched legacy of past injustice for Indigenous people and Dalits remains.

Disaster's Unequal Effects

The 2015 earthquake caused colossal damage to life, property, livelihood, and environment across the region. But, there were substantial variations in damage between geographic location, class, caste, ethnicity, and gender (He et al. 2018; Sujakhu et al. 2019). In particular, Tamang and Dalits were disproportionately impacted by the disaster with ramifications for longer term worse outcomes. Noting the heavy toll on Tamangs, Gaha Magar (2072 BS) writes of "Tamang in the epicenter" of the earthquake. Ghale (2016) makes a similar point. Pointing to historical dispossession and powerlessness from the Gorkha conquest in the late 18th century, Holmberg and March (2015) termed the devastating disaster in the Tamang territory as a "Tamsaling tragedy."[1] Campbell emphasized the Tamang peoples' 'culturally distanced relationship to the state' symbolizing 'a population marked by centuries of structural violence' (Campbell 2018: 111).

Based on their research the 2015 earthquake in Dalit communities in Sindhupalchok District, Bownas and Bishokarma (2019) show how the weight of past 'traditional' discrimination shaped the present differentiation of outcomes by caste. Dalits, they show, were slightly overrepresented as victims compared to their population. Dalit homes often located in hazardous locations were built of mud and prone to earthquakes. Landlessness compounded shelter and livelihood problems (FEDO 2017). Despite the fact that Dalits face systematic exclusion, they are often invisible in disaster discourse. Such silence, Folmar (2015) suggests, reaffirms caste discrimination in Nepal.

Our data suggest that Tamang and Dalit suffered disproportionately from the disaster. First, they died in disproportionate numbers. Tamangs constitute 20.7% of the 14 most affected districts' population but 37.2% of those who died. Conversely, hill Brahman and Chhetris, who constitute 36.1% of the population, saw a death rate of only 20.1%. The Dalit castes – Kami, Damai/Dhole, and Sarki – constitute 6% of the affected region's population but 7.7% of those who died. Gurungs were also overrepresented in earthquake deaths. The primary reasons for relatively higher deaths in marginalized groups are poverty, less sturdy house construction, and living in locations that were more vulnerable to natural hazards (Figure 12.1).

The second disparity in the impact is seen in the number of physical injuries. Of the total 2,419 persons in sample who reported to have suffered physical injury during the earthquake, the highest are from among the Tamang population with

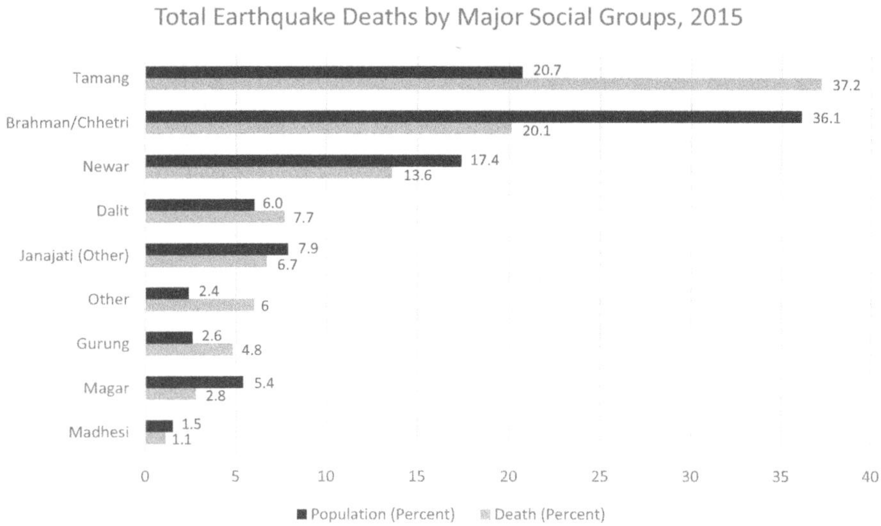

Figure 12.1 Percentage of deaths by ethnic group from the 2015 Earthquake relative to total population.

Sources: MOHA 2015, CBS 2011. Tamang 2019. Figure credit: author.

4.3% of injuries. This is more than double compared to the average (1.9%) for the overall affected population. In addition to death, the physical injury of the family member, especially if the injured is the main earning person, the deleterious effect on family wellbeing is very high.

The third impact is with psychosocial illness. Among the sample population, the highest percent of the people who continue to experience psychosocial problems was among Tamangs. The psychosocial problems include depression, anxiety, stress, and other mental health issues. These problems are caused by disruptions in a personal life through grief, trauma, and loss. Further, the stress of acquiring adequate shelter, financial matters, and rehabilitation of livelihood are other factors that causes mental health issues. Other subtler impacts such as problems of adjustment in post-disaster situations and disruption of spiritual practice also factor behind these impacts.

Uneven Recovery and Current Household Capacity

Governments can earn legitimacy by helping citizens during emergencies. Through the equitable distribution of materials, disaster recovery work can reduce inequality and attempt to redress past injustices. Disaster reconstruction and recovery, however, in Nepal's case, was a missed opportunity for achieving social equity.

Disaster recovery depends on addressing vulnerabilities. Various studies show that vulnerable populations face longer recoveries and worse outcomes. Several factors cause disparities: asymmetric power relations (Lam 2023), unequal physical and financial capacities and access to resources (Rawal 2021), context and location (Spoon 2020), and accountability failure of the government (Shrestha et al. 2019). The case from Sindhupalchowk shows how progress differed across gender, caste/ethnic as well as income groups (Sapkota et al. 2021). Examining Nepal, Amnesty International (2017) underscored the role of state-led reconstruction works in exacerbating processes of marginalization. All this indicates that injustice existed in the past, but also in the present.

Our research data also shows that the reconstruction and recovery for Tamang and Dalits have been lower than those for better-off groups. As of June 2019, to take one important measure, only 66% of poorer households had built a new house compared to 77.6% of richer households. Seventeen percent of families still lived in temporary shelters, such as mud and wood sheds with corrugated zinc roofs, or part of a damaged house.

Per-capita expenditure also showed a disparity for Tamangs and Dalits. The overall average per-capita annual household expenditure in 2019, four years after the earthquake, was NRs. 73,110. Brahmin (NR. 91,270) and Newar households (NRs. 89,148) were far above average. On the other end, Dalits averaged NRs. 59,418 and Tamangs NRs. 64,119.

A composite index of the recovery status[2] based on four primary indicators – status of house reconstruction, number of the three tranches of grants taken, level of per capita expenditure, and overall bounced-back perception – reveals that Brahmins have achieved considerable success with 83.7% recovery. Dalits (62.3%) and Tamangs (63.5%) remained far behind in terms of overall recovery status. Indebtedness was an unintended consequence from which Tamang and Dalit continued to suffer. Many households from vulnerable communities could only build housing units (even with inadequate space) mostly by taking high-interest loan to supplement government grants. Three out of five Tamang families had to take loans for house reconstruction or to meet family basic needs. Dalits faced a similar situation. Disaster also forced both to divert resources from investment in family education and health.

Our Poverty Probability Index (PPI),[3] integrates measures of access to education, information, and technology, basic social service, social capital, remittance, disaster preparedness awareness, school dropout rate, illness, wage employment, and demographic pressure. PPI results revealed an uneven distribution. Newars (74.4%) and Brahmins (70.3%) possessed the highest household capacity, while Tamangs (53.8%) and Dalits (51.9%) had the lowest.

Examining just poverty indicators, our research showed that in the 14 earthquake-affected districts 15.1% of households were likely to be below the poverty line. Tamangs had the highest percentage of households (19.1%) in poverty. Newars had the lowest percentage of households below poverty (9.9%) followed by Brahmins (11.8%).

Dalits and Tamangs had significantly less “linking” or “bridging” social capital. Linking social capital means access to formal institutions, including the state, and connections to people with influence (Granovetter 1973). Over 45% of Brahmin households reported knowing at least one influential person. Only about one-fourth of Tamangs and Dalits reported a link with an influential person.

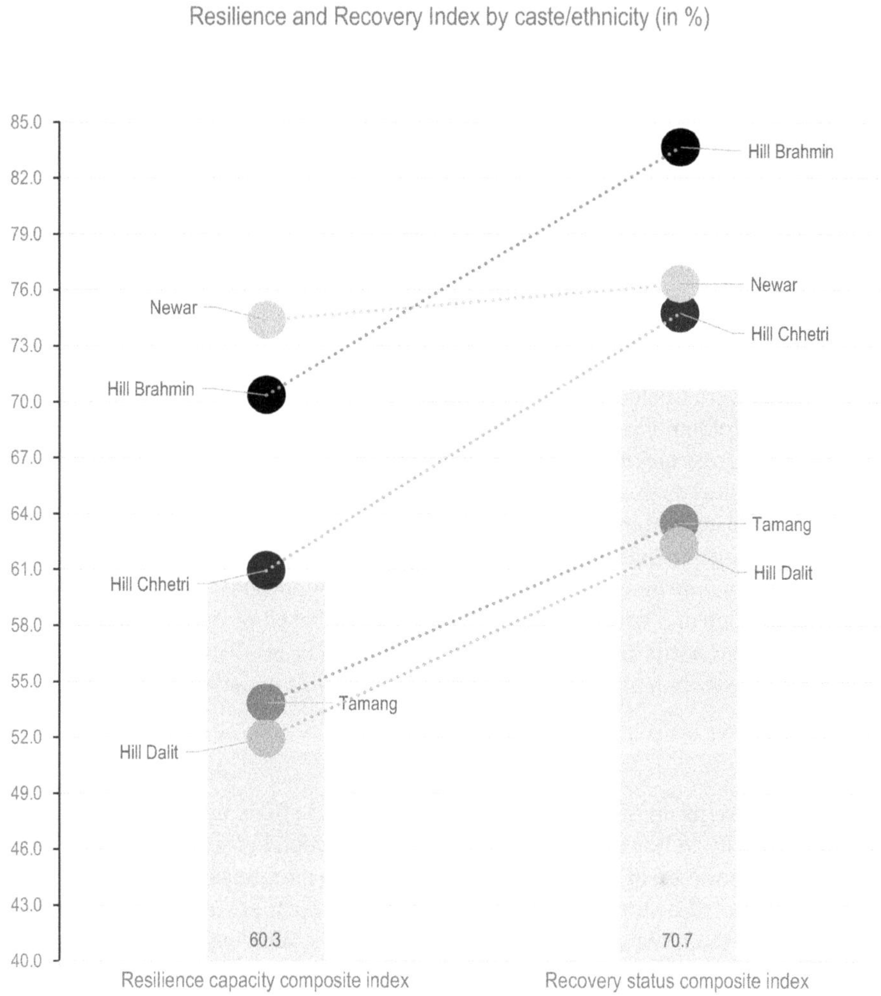

Figure 12.2 Household capacity and recovery status composite index by social group.

Source: Tamang 2019. Figure credit: author.

A clear correlation existed between household capacity and recovery status. A higher household capacity generally meant better recovery through better access to government resources and private capital. As Figure 12.2 shows, Brahmin, Newar, and Chhetri households with higher capacity achieved higher levels of recovery, while Dalit and Tamang score lower bottom in both variables. Dalit and Tamang scores dropped below average both in household capacity (60.3%) and recovery rate (70.7%).

The uneven recovery after the 2015 earthquakes shows a failure on the part of the Nepali government to reduce inequality through equitable and inclusive measures. There were three major reasons.

The first was disregarding restorative justice. The government took the usual approach of giving "equal amount to all victims" according to the general principle of equal treatment. But an equal treatment approach that does not recognize the specific needs of society's marginalized and vulnerable usually results in unequal outcomes; it can even exacerbate pre-existing inequalities. The government disregarded the need for special measures or affirmative action required for the marginalized.

The second was exclusionary disaster governance. In reconstruction discussion and mechanisms, decision-making was top-down. The NRA embodied centralized authority that operated through civil servants. International donors and the bureaucrats defined the recovery and carried out recovery effort without the active participation of the stakeholders in decision-making and implementation. Local communities were treated merely as recipients.

The third problem was non-recognition of Indigenous and local knowledge and heritage. The government's approach to reconstruction focused on technical and financial measures such as housing design and grants. The government prescribed designs for houses without consultation with owners. Although the new structures are more earthquake-safe, government design control undercut practical requirements, traditional design, and aesthetics, and more importantly, it undercut Indigenous knowledge and heritage. Recovery responses based on limited consultation with and limited participation of affected people led to ineffective programs, disempowerment of communities, and perpetuation of past injustice.

Conclusion

This paper presents empirical evidence of the unequal effect, uneven recovery status, and disparity in household capacity of different social groups in Nepal's 2015 earthquakes. Because of the past injustice, the 2015 earthquakes have disproportionately impacted historically marginalized groups such as Tamangs and Dalits. The impact of the disaster in terms of deaths, injuries, and loss of livelihoods has been unequal. The recovery process has also been uneven: some have bounced back better while others have fallen behind compared to before the disaster. In some cases, the recovery effort even exacerbated inequalities.

The findings highlight historical social inequities and the failure of post-disaster interventions to address these inequalities. A longitudinal view that takes into account both pre- and post-disaster situations is helpful to put disaster justice from a social, political, and economic perspective. With a new Constitution adopted amidst disaster that promises the right to a clean and safe environment, health, food, housing, and education as equal rights of the citizens, Nepal has a great opportunity to achieve environmental and disaster justice.

Acknowledgments

This paper is based on the data collected through research on Community Resilience Capacity in the context of the 2015 earthquakes in Nepal through the Central

Department of Anthropology at Tribhuvan University with funding assistance from USAID. I thank Yogendra Gurung and Mohan Khajum for their assistance in the analysis of the survey data. Special thanks to Tom Robertson, Jonathan London, and Jagannath Adhikari for their valuable review comments on the draft paper.

Notes

1 Tamsaling is the name of the province proposed by the Committee for State Restructuring of the Constitution Assembly elected in 2008 covering the ten districts surrounding Kathmandu valley in recognition of the historical identity of Tamang people. The districts include Chitwan, Dhading, Dolakha, Kavrepalanchowk, Makawanpur, Nuwakot, Ramechhap, Rasuwa, Sindhuli, and Sindhupalchowk.

2 The composite index for recovery status was computed as a simple average of the variables (house constructed, all three grants received, overall bounced-back perception, and expenditure per capita). The amount of expenditure was normalized in terms of 100. The formula [(Actual - Minimum)/ (Maximum - Minimum)] *100 was used for calculation.

3 See Schreiner (2013) for the Poverty Probability Index (PPI) as a poverty measurement tool.

References

Adhikari, J., Rai, M. K., Subedi, M., & Baral, C. (2023). Foreign Labour Migration in Nepal in Relation to COVID-19: Analysis of Migrants' Aspirations, Policy Response and Policy Gaps from Disaster Justice Perspective. *Journal of Ethnic and Migration Studies*. https://doi.org/10.1080/1369183X.2024.2268983.

Amnesty International (2017). *Building Inequality: The Failure of the Nepali Government to Protect the Marginalised in Post-earthquake Reconstruction Efforts*. London: Amnesty International. Retrieved from www.amnesty.org/download/Documents/ASA3160712017ENGLISH.PDF.

Bankoff, G. (2018). Blame, Responsibility and Agency: 'Disaster Justice' and the State in the Philippines. *Environment and Planning E: Nature and Space*, 1(3), 363–381.

Bennett, L., & Parajuli, D. (2012). *The Nepal Multidimensional Exclusion Index: Making Smaller Social Groups Visible and Providing a Baseline for Tracking Results on Social Inclusion*. Washington, DC: The World Bank.

Bhattachan, K., Sunar, T., & Bhattachan, Y. (2009). *Caste Based Discrimination in Nepal*. Working Paper Series Vol. III, No 8. New Delhi: Indian Institute of Dalit Studies.

Bhattarai, A. M. (2018). Earthquake and Earth Justice: Emergence of the Environmental Justice Movement and Its Relevance in Addressing Unanticipated Events. In J. H. Kruhl, R. Adhikari, & U. E. Dorka (eds.), *Living under the Threat of Earthquakes* (pp. 285–292). New York: Springer.

Bownas, R., & Bishokarma, R. (2019). Access after the Earthquake: The Micro Politics of Recovery and Reconstruction in Sindhupalchowk District, Nepal, with Particular Reference to Caste. *Contemporary South Asia*, 27(2), 179–195.

Campbell, B. (2018). Communities in the Aftermath of Nepal's Earthquake. In L. Bracken, H. A. Ruszczyk, & T. Robinson (eds.), *Evolving Narratives of Hazard and Risk: The Gorkha Earthquake, Nepal 2015* (pp. 109–123). New York: Palgrave Macmillan.

CBS. (2014). *National Population and Housing Census 2011, Social Characteristics*. Kathmandu: Central Bureau of Statistics, National Planning Commission, Government of Nepal.

Cutter, S. L. (2006). The Geography of Social Vulnerability: Race, Class and Catastrophe. In *Understanding Katrina: Perspectives from the Social Sciences*. New York: Social Science Research Council.

Douglass, M., & Miller, M. A. (2018). Disaster Justice in Asia's Urbanizing Anthropocene. *Environment and Planning E: Nature and Space*, 1(3), 271–287.

Eda, H. (2015). Disaster Justice: Mobilising Grassroots Knowledge against Disaster Nationalism in Japan. In A. J. Jolivette (ed.), *Research Justice: Methodologies for Social Change* (pp. 95–106). Bristol: Policy Press.

FEDO. (2017). *Impact of the Earthquake: Access of Dalits to the Recovery and Reconstruction*. Kathmandu: Feminist Dalit Organization (FEDO).

Flacke, J., Maharjan, B., Shrestha, R., & Martinez, J. (2022). Environmental Inequalities in Kathmandu, Nepal: Household Perceptions of Changes between 2013 and 2021. *Frontiers in Sustainable Cities*, 4, 1–19.

Folmar, S., Cameron, M., & Pariyar, M. (2015). Digging for Dalit: Social Justice and an Inclusive Anthropology of Nepal. Aftershocked: Reflections on the 2015 Earthquakes in Nepal. Retrieved from https://culanth.org/fieldsights/digging-for-dalit-social-justice-and-an-inclusive-anthropology-of-nepal.

Gaha Magar, S. (2015 [2072 BS]). *Punnirman: Kendra Binduma Tamang* (Reconstruction: Tamang in Epicenter). Himal Khabar Patrika, 5–11 July 2015. Retrieved from http://nepalihimal.com/article/4607.

Ghale, S. (2016). A Year After Nepal's Killer Quake, the Tamangs Continue to Struggle on the Margins'. *The Wire*. Retrieved from http://thewire.in/2016/05/22/ayearafternepal earthquaketamangcommunitycontinuestostruggle37305/ [accessed 5 May 2016].

Granovetter, M. S. (1973). The Strength of Weak Ties. *American Journal of Sociology*, 78(6), 1360–1380.

He, L., Aitchison, J. C., Hussey, K., Wei, Y., & Lo, A. (2018). Accumulation of Vulnerabilities in the Aftermath of the 2015 Nepal Earthquake: Household Displacement, Livelihood Changes and Recovery Challenges. *International Journal of Disaster Risk Reduction*, 31, 68–75.

Höfer, A. (2004 [1979]). *The Caste Hierarchy and the State in Nepal. A Study of the Muluki Ain of 1854. Innsbruck: Universitatsverlag Wagner*. 2nd Ed. Lalitpur: Himal Books.

Holmberg, D., & March, K. (2015). Tamsaling and the Toll of the Gorkha Earthquake. Aftershocked: Reflections on the 2015 Earthquakes in Nepal. Retrieved from https://culanth.org/fieldsights/tamsaling-and-the-toll-of-the-gorkha-earthquake

Holmberg, D. H. (1999). Local Production/Local Knowledge: Forced Labour from Below (co-authored with Kathryn March and Suryaman Tamang). *Studies in Nepali History and Society*, 4(1), 5–64.

Lam, M. L. (2023). Asymmetrical Power Relationships in Disaster Reduction Remain the Key Barrier to Sustainable Recovery: Neoliberalism and Nepal's Post-Earthquake Recovery Experiences. *Sustainable Development*, 1–13. https://doi.org/10.1002/sd.2683.

Lawoti, M. (2008). Exclusionary Democratization in Nepal, 1990–2002. *Democratization*, 15(2), 363–385.

NPC. (2015). *Post Disaster Needs Assessment (PDNA)*. Kathmandu: National Planning Commission.

NRA. (2018). *Rastriya Punnirman Pradhikaran, (2075 BS), Aarthik Barsha 2074/75 ko Barshik Pratibedan* (National Reconstruction Authority, [2018] Annual Progress Report of Fiscal Year 2017/2018). Kathmandu: Government of Nepal.

NSAC. (1998). *Nepal Humand Development Report 1998*. Kathmandu: Nepal South Asia Centre.

Rawal, V., Bothara, J., Pradhan, P., Narasimhan, R., & Singh, V. (2021). Inclusion of the Poor and Vulnerable: Learning from Post-Earthquake Housing Reconstruction in Nepal. *Progress in Disaster Science*, 10, 100162.

Rawls, J. (1971). *A Theory of Justice*. Cambridge: Harvard University Press.

Reid, M. (2013). Disasters and Social Inequalities. *Sociology Compass*, 7(11), 984–997.

Riaz, A., & Basu, S. (2007). *Paradise Lost? State Failure in Nepal*. Pennsylvania: Lexington Books.

Robertson, T. (2017). Unnatural Disasters. *My Republica*. Retrieved from https://myrepublica.nagariknetwork.com/news/unnatural-disasters/ [accessed 15 October 2017].

Sapkota, J. B., Kurita, K., & Neupane, P. (2021). Progress after the 2015 Nepal Earthquake: Evidence from Two Household Surveys in One of the Hardest-Hit Mountain Villages. *Sustainability*, 13, 11677. https://doi.org/10.3390/su132111677.

Schreiner, M. (2013). Simple Poverty Scorecard for Poverty Assessment Tool, Nepal. Retrieved from www.simplepovertyscorecard.com.

Shrestha, K. K., Bhattarai, B., Ojha, H. R., & Bajracharya, A. (2019). Disaster Justice in Nepal's Earthquake Recovery. *International Journal of Disaster Risk Reduction*, 33, 207–216.

Spoon, J., Gerkey, D., Chhetri, R. B., Rai, A., & Basnet, U. (2020). Navigating Multidimensional Household Recoveries Following the 2015 Nepal Earthquakes. *World Development*, 135. https://doi.org/10.1016/j.worlddev.2020.105041.

Sujakhu, N. M., Ranjitkar, S., He, J., Schmidt-Vogt, D., Su, Y., & Xu, J. (2019). Assessing the Livelihood Vulnerability of Rural Indigenous Households to Climate Changes in Central Nepal, Himalaya. *Sustainability*, 11, 2977. https://doi.org/10.3390/su11102977.

Tamang, M. S. (2009). Tamang Activism, History and Territorial Consciousness. In D. N. Gellner (Ed.), *Ethnic Activism and Civil Society in South Asia* (pp. 269–290). New Delhi and London: SAGE.

Tamang, M. S. (2020). *Community Resilience Capacity: A Study on Nepal's 2015 Earthquakes and Aftermath*. Kirtipur: Central Department of Anthropology, Tribhuvan University.

Tamang, P. (1992). *The Tamangs: A Face of Nepal*. Kathmandu: Nepal Tamang Ghedung.

Verchick, R. (2012). Disaster Justice: The Geography of Human Capability. *Duke Environmental Law and Policy Forum*, 23, 23–71.

13 Indigenous Struggles for Development Justice in Nepal

Environmentalism on the Ground

Prabindra Shakya

Introduction

Indigenous Peoples in Nepal, like those around the world, have been the stewards of the lands, forests, rivers, and other natural resources of the country for millennia. However, in the process of nation-building, they have gradually lost ownership and control over their lands and resources in the name of both "conservation" and "development." Subsequently, there has been significant harm to the environment that not only affects Indigenous Peoples but also the wider Nepali society.

A case in point is the "fortress conservation" model primarily followed in Nepal, which has resulted in continuing harm for Indigenous and other local communities (McLean 1999; Mishra 2004; Müller-Böker 1991; Stevens 1993). Over the last 50 years, national parks, conservation areas, and hunting reserves have been set up in Indigenous territories relying upon the army and a model of militarization for security (Robertson 2018; Thing 2019). As a result, the Indigenous communities that once stewarded those forests and ecosystems have been displaced from their lands. They have been denied access to their resources, ignoring or devaluing their histories of sustainable resource management. Community members have been subjected to harassment, detention, fines, abuse, torture, and even killings at the hands of military and forest personnel while Indigenous women have also been raped and sexually abused (Ghale 2018).

The ongoing aggressive drive for economic growth has also dispossessed many Indigenous communities. They have been displaced from their houses, lands, and cultural and other sacred sites for the construction of dams, roads, transmission lines and other infrastructure projects. In return, they receive few benefits, even as political and business elites cash in.

Ill-planned "development" projects – often termed as projects of "national pride" – that harm the affected Indigenous communities also cause disproportionate wider impacts on the natural and cultural environment. Opposition of the Indigenous communities to those projects is considered anti-national, anti-development, ethnically divisive, or driven by narrow political or financial interests while the environmental impacts are deemed limited to the project areas. On the contrary, Indigenous Peoples in Nepal have been fighting back to defend their

DOI: 10.4324/9781003371175-17
This chapter has been made available under a CC-BY-NC-ND 4.0 license.

rights, communities, cultures, identities as well as the environment (Dhungana and Maskey 2019; Sikor et al. 2019).

Unlike in other parts of the world, Indigenous communities' struggles in Nepal receive little support from environmentalists and environmental justice advocates. This chapter argues that their movements are not just challenging social and development injustice against them but are also aimed at promoting environmental justice for the wider society and the country. This chapter, based on real-life practices of advocacy for justice, will present the stories of struggles of Indigenous communities across the country that depict the "environmentalism of the poor" (Martinez-Alier 2003) in a true sense. The chapter will be grounded in the understanding of the environment as a holistic system in which Indigenous communities and their agency are an indispensable part.

Indigenous Peoples, Environment and Development in Nepal

Nepal's Constitution refers to the country as a "multi-ethnic, multi-lingual, multi-religious, multi-cultural" nation (Government of Nepal 2015). Despite a relatively small area – about 147,500 sq km, Nepal is home to 60 officially recognized Indigenous Peoples, who are referred to as Adivasi Janajati (Indigenous nationalities).[1] They are legally defined as "tribe[s] or communit[ies] [...] having [their] own mother language and traditional rites and customs, distinct cultural identity, distinct social structure and written or unwritten history" (Government of Nepal 2002). They are categorized based on their human development as either endangered, highly marginalized, marginalized, disadvantaged, or advanced groups. They are also identified based on their original settlements as mountain, hill, inner Terai, and Terai groups.

According to the 2011 national census, Indigenous Peoples account for 36% of Nepal's around 30 million population. However, Indigenous Peoples' organizations claim their number could be as high as 50%[2] while many groups that identify as Indigenous Peoples also await official recognition. The population share of Indigenous Peoples as per the recent 2021 census is yet to be ascertained as the census figures themselves are disputed. About 110 of the 123 languages spoken in the country are in use among Indigenous communities.[3] Similarly, they represent significant cultural diversity that is rooted in their relationship with and traditional knowledge of their lands, territories and resources.

Eighteen of the twenty protected areas of the country, including twelve national parks, covering about 23% of the national land area, are in the ancestral domains claimed by various Indigenous groups. Studies show that the benefits of securing Indigenous Peoples' rights to their lands, territories, and resources extend to protecting the wider environment and combating climate change, and that investing in the security of Indigenous forestlands reduces deforestation and represents a low-cost, high-benefit approach to climate change mitigation with huge environmental and economic benefits.[4] However, the existing legal framework of Nepal does not adequately recognize the customary rights of Indigenous Peoples over forest or other resources.[5]

Nepal ratified the Indigenous and Tribal Peoples Convention of the International Labor Organization (ILO), 1989 (Convention 169) in 2007 – to date being the only Asian country to do so.[6] Same year, Nepal also voted in favor of adoption of the groundbreaking United Nations Declaration on the Rights of Indigenous Peoples (UNDRIP) at the UN General Assembly. ILO Convention 169 and UNDRIP obliges Nepal to recognize and protect the rights of Indigenous Peoples, including the lands, territories, and resources that they traditionally occupy or use, and to determine their priorities and strategies for development. It is also required to obtain the free, prior, and informed consent (FPIC) of the concerned Indigenous Peoples for any activities undertaken on their lands. FPIC allows the Indigenous communities to negotiate the conditions under which such projects will be designed, implemented, monitored, and evaluated and also includes their right to say no to projects they consider harmful. Nevertheless, the human rights obligations of Nepal under those instruments, including obtaining FPIC, still need to be fully internalized in the national legal system – not to mention their application on the ground.

Nepal's Constitution requires the State to obtain the participation of Indigenous nationalities in decisions concerning them by making special provisions to ensure their right to live with dignity, along with their identity, and protect and promote their traditional knowledge, skill, culture, social traditions, and experiences.[7] The current Five-Year Plan of the Government includes provisions promoting the inclusion and participation of Indigenous and local communities in the development process of the country, and in the management of natural resources.[8] The situation on the ground, however, shows such inclusion and participation are far from reality.

Indigenous Peoples' Struggles against Development Injustice and for Environmental Justice

Infrastructure and other "development" projects in Nepal are often carried out without meaningful consultation with the affected Indigenous communities – let alone obtaining their FPIC. These include large-scale hydropower projects and ill-planned infrastructure construction with disproportionate impacts on the communities in the form of dispossession and displacement from their lands and resources receiving little compensation or benefits. On the other hand, the harms caused by those projects on the natural and cultural environment are not even given due consideration while the opposition of the affected communities is interpreted in narrow financial terms.

Impacts of Hydropower Development and Majhi Resistance

Harms of large-scale hydropower projects on Indigenous communities and the wider environment and opposition of the communities against such projects are evident in the case of cascade of dams being constructed along the Sunkoshi River in eastern Nepal. The river is an important source of livelihood and identity for the many Indigenous Majhi communities living along the river. One of the major rivers of Nepal, Sunkoshi is one of the seven tributaries of the Koshi River (also known as Saptakoshi) – a transboundary river that flows through China, Nepal, and India. For the Majhis, who are traditionally fisherfolk, the river is not only important for their livelihoods of fishing and boating but also holds a spiritual value – they refer to the river as their "uncle."

The Majhis carry out annual worship of rivers (called the Koshi Puja or Ladi Puja) on its banks or at the confluence of the river and its tributary rivers and streams. Their traditions require the flowing water of a river or stream for their death rites. They set afloat on the river the *Khaula,* a leaf bouquet, created in the name of the dead with the belief that the departed soul will flow away to heaven. In the Dudhkoshi River in far-eastern Nepal (another tributary of the Saptakoshi), Majhis even discharge the dead body itself in the river. They also need fish and other resources from the river for other life events such as birth and marriage.

In recent times, due to depletion of fishing populations in the river, including because of the Koshi Barrage Dam constructed between 1952 and 1968 near the southern border of Nepal with India, much of the younger generations of Majhi have left their home villages for work elsewhere. Currently, three dams are at different stages of development under Sunkoshi-Marin Diversion Multipurpose Project and Sunkoshi-2 and Sunkoshi-3 hydropower projects – the latter two are mega dams with large reservoirs. All those projects will have disproportionate impacts on the Majhi communities while also adversely affecting the river's ecosystems. Consequently, all three projects have raised various concerns among the affected Majhi communities.[9]

In 2021, the Majhi communities in Ramechhap district affected by the Sunkoshi Marin Diversion Multipurpose Project – an irrigation-cum-hydropower project with a smaller dam of 12 meters and inundation area of 350 hectares – raised various concerns and demands about the project. Those include calls for information disclosure and meaningful consultations about the project to obtain their FPIC on matters concerning them; fair and adequate compensation for their lands and houses that will be inundated; as well as preservation of their cultural sites and traditions that will be affected by the project.[10] Following additional consultations by the project with the affected communities, the Government has promised higher compensation than originally offered at the time of its impact assessment and community development programs have been proposed with the project budget. The project is currently under construction but the Majhi populations are concerned whether the terms agreed will be met.

Similar demands have also been made during the public hearings organized for the environmental impact assessment of the 536MW Sunkoshi-3 hydropower project, which will export power to Bangladesh. The Government has recently approved the assessment report for the project planned for construction on the border of Ramechhap and Kavrepalanchok districts. According to the environmental impact assessment, the project will also affect areas along the river in Sidhupalchok and Sindhuli districts. Approximately 4,500 families will lose their land due to project land acquisition. Of them, more than 1,600 families will see their homes and other buildings inundated.

Beyond this, the affected Majhi communities have called for outright scrapping of the Sunkoshi-2 Hydropower Project, the largest among the cascade dam projects. The 978MW project planned on the border of Ramechhap and Sindhuli districts will inundate an area of 4,500 hectares up to at least 53 km from the 16 m high dam. As a result, a majority of settlements (most of those ancestral to the Majhis) along the riverbanks will be submerged, displacing around 6,000 households and impacting 40,000 people. According to an Asian Development Bank (ADB) study, there are about 800 fishing (Majhi) households within the project's dam and inundation area whose livelihood depend on fishing in Sunkoshi, Tamakoshi, and their tributaries; however, the numbers are likely be even greater due to challenges in enumerating informal economies.[11]

According to the Majhis, the Sunkoshi-2 hydropower project, along with the other projects, will result in an "irreparable impact on the natural life of the river and result in the disappearance of the culture, identity, and settlement of the Indigenous Majhi associated with the river and their civilization developed along the River."[12] The affected communities have thus disrupted the public hearings for the environmental impact assessment of the projects and called on the national authorities and international stakeholders, including the Asian Development Bank (ADB), to cancel them. Clad in their indigenous attire, playing their traditional drums and other musical instruments, carrying fish- and boat-shaped banners, and throwing fish nets in the streets, they have demonstrated by the hundreds at the local levels to make their demand for the annulment of the projects heard.

The Majhi leaders and elders have asserted that no amount of compensation can redress the significant losses they would suffer under the projects. The impacts of the project on the cultural life of Indigenous communities, on their religious sites, on their livelihoods and on the wider environment are far greater than the benefits they would get from them, given that the electricity produced by the various dams will be most likely be deployed to Kathmandu Valley or exported from Nepal. At the same time, they have called on the Government to assess the cumulative impacts of the cascade of dams planned along the Sunkoshi River.

These are some of the ways in which the Majhi communities are struggling to save their culture and protect the environment, which are interconnected in a relationship developed over the centuries.

Newa Opposition to Destructive Expressway Construction

For decades, Indigenous Newa communities in Lalitpur district in the south of Kathmandu Valley have been fighting back against various infrastructure projects to protect their culture and the environment. Key among them is the first expressway of Nepal undergoing construction – the Kathmandu-Terai/Madhesh Fast Track highway, which is sponsored by the Government and being implemented by the

Nepali Army since 2017.[13] For the construction of the expressway, the Government has called for large-scale acquisition of the farmlands of hundreds of families as well as the communal and religious lands of Newa people in Bungamati and Khokana, which will result in the displacement as well as disintegration of their communities and identities.[14]

The farmlands in the project area are among the last remaining major swathes of agricultural areas in Kathmandu Valley which once was self-reliant in agricultural produce due to its very fertile topsoil. Furthermore, the communal (*Guthi*) lands include farmlands that sustain various religious temples and fairs or festivals, open fields and sites used for traditional dances and death rites, as well as areas that are believed to hold the archaeological remains of the earliest settlements of Newa people. As stated by an activist in Khokana in a news report in Nepali Times, "Our land is us and we are our land. People here have a physical, spiritual, social, cultural and economic connection with the land. We have no identity without it."[15]

The Newa communities have opposed the construction of the Fast Track expressway on their lands since it was proposed. The communities' opposition was noted even in the environmental impact assessment (EIA) report of the project submitted to the Government in 2015 after its feasibility study was undertaken with assistance from the ADB. During public hearings, there was significant objection to government acquisition of productive agricultural land near Khokana, the local population had already sacrificed enough land for other projects. The EIA consultants noted that an alternative alignment of the expressway along the west bank of the Bagmati River as far south as 9 kilometers would pass through severely degraded areas well away from Khokana.[16] However, the Government turned deaf ears to the communities' discontent and the EIA suggestions for alternatives.

As a result, the communities have been mobilizing year after year calling for realignment of the Expressway to move its entry point further south where it can connect with other highways entering the valley. During their protests, they have faced retaliation in various forms and even had clashes with police during which tear gas and live bullets have been fired, injuring many. In one such clash in July 2020, around the time of the annual rice planting, over a dozen protestors were injured by the military. The Government has placed a Nepal Army camp on the land forcibly acquired for the expressway. The militarization has caused fear in the local communities.[17]

Besides the expressway, many other projects are also planned in Khokana and Bungamati, including the Kathmandu Outer Ring Road, Bagmati River Basin Improvement Project (Bagmati Corridor), Thankot-Bhaktapur Transmission Line Project and one of the four "Smart Cities" proposed in Kathmandu Valley – some of which are stalled or have been suspended due to opposition of the communities, and all directly or indirectly supported by the ADB. The leaders of these communities assert that all those projects, if implemented, will entirely displace the agrarian Newa communities from their ancestral lands and settlements. At the same time, it will also wipe out the last remaining green fields and open spaces in the south of Kathmandu Valley, which is being choked due to worsening air pollution and uncontrolled population growth.

The affected Newa communities in Khokana and Bungamati have called for help from Nepal's Supreme Court and the UN human rights agencies for the defense of their rights and their lands against the projects.[18] Nonetheless, the Government is seemingly trying to push through the projects with brute force.

The Government has used force in other places as well. For example, in the town of Kwena (Bhutkhel) in Kirtipur, the Government recently rammed through the construction of a World Bank financed Chobhar dry port with the help of security forces, despite the resistance of the affected Indigenous Newa people and other locals.[19] They had opposed the construction, demanding the return of their lands on which the dry port has been built as well as owing to the impacts of the dry port on nearby cultural and environmental resources. The land had earlier been acquired for a cement factory that is now shut. After the closure of the factory in 2002, locals wanted their lands to be used for social purposes such as the construction of a hospital or in line with their aspiration to develop the area as a tourism site, for which they had even constructed a locally managed park nearby. The dry port that was inaugurated in 2022 amid heavy security has attracted few traders raising questions over its viability – an issue that was raised by those affected throughout their opposition to the project.

Need for a Paradigm Shift for Environmental Justice in Nepal

There needs to be a paradigm shift towards recognition, protection, and promotion of the rights of Indigenous Peoples in Nepal. Without this, environmental justice will never be achieved in the country. Indigenous Peoples here have been practicing environmentalism on the ground for centuries and their efforts should be supported and advanced further. Indigenous Peoples should not have to carry a disproportionate burden due to present-day unsustainable development spree.

With the ongoing climate and other crises facing humanity, it is even more important to assert the fundamental role played by Indigenous Peoples to protect the environment, for which securing their lands and resources is prerequisite. Indigenous Peoples possess invaluable knowledge about coexisting in harmony with nature, using nature's resources without exploiting them, benefiting from nature's offers without destroying it.

The development projects need not be delayed or suspended if implemented with the consent of the concerned Indigenous Peoples of the land and they take nature into account. Nepal's Indigenous People are not – as they have been sometimes referred to – "anti-development"; they just want a more just, fair, environmentally friendly development based on the needs of the communities and of the lands they live on, and not imposed from the authorities, international lenders, and businesses far away. Unfortunately, when Indigenous People assert their rights and raise their voices about unfair development, they face reprisals, even deadly attacks.[20]

As an Indigenous Kichwa youth activist from Ecuador said, "We [Indigenous Peoples] have to be made visible to the world to say: 'We are here, we exist, and we are protecting the environment for everyone, not just for us."[21]

But as it is, Indigenous Peoples in Nepal are struggling for their existence in the face of a destructive infrastructure-building spree all in the name of development.

Notes

1 The National Foundation for Development of Indigenous Nationalities (NFDIN) Act lists 67 Indigenous nationalities. https://nfdin.gov.np/eng/page/caste-and-ethnicity, accessed October 30, 2023.

2 International Work Group for Indigenous Affairs. (2022). *The Indigenous World 2022*, p. 266.

3 Indigenous Voice. (2023). Indigenous Peoples of Nepal. https://english.indigenousvoice.com/indigenous-peoples-of-nepal, accessed November 21, 2023.

4 World Resources Institute Blog. (2017). *By the Numbers: Indigenous and Community Land Rights.* World Resources Institute. (2016). *Climate Benefits, Tenure Costs: The Economic Case for Securing Indigenous Land Rights in the Amazon.*

5 Government of Nepal: Ministry of Forests and Environment. (2018). Chapter 3. *Nepal National REDD+ Strategy* (p. 11).

6 International Labor Organization. (1989). *Ratifications of C169 - Indigenous and Tribal Peoples Convention (No. 169).*

7 Constitution of Nepal (2015). Art. 51 § (j)(8).

8 Government of Nepal: National Planning Commission. (2020). *The Fifteenth Plan (Fiscal Year 2019/20–2023/24).*

9 *Nepal's indigenous Majhi communities demand cancellation of Sunkoshi-2 Hydropower Project.* (October 6, 2021). Community Empowerment and Social Justice Network. https://cemsoj.wordpress.com/2021/10/06/nepals-indigenous-majhi-communities-demand-cancellation-of-sunkoshi-2-hydropower-project/, accessed December 14, 2023.

10 *Indigenous Majhi and Local Communities Affected by Sunkoshi-Marin Diversion Multipurpose Project Submit 13-Point Demands to the Local Authorities* (January 18, 2021). Community Empowerment and Social Justice Network. https://cemsoj.wordpress.com/2021/01/18/indigenous-majhi-and-local-communities-affected-by-sunkoshi-marin-diversion-multipurpose-project-submit-13-point-demands-to-the-local-authorities/, accessed December 14, 2023. See also Right Energy Partnership. (April 26, 2021). *Indigenous Majhi on the Line between Development and Existence.*

11 Asian Development Bank. (2013). *Nepal: Project Preparatory Facility for Energy*, Section C, para. 52.

12 *Nepal's Indigenous Majhi Communities Demand Cancellation of Sunkoshi-2 Hydropower Project* (October 6, 2021). Community Empowerment and Social Justice Network. https://cemsoj.wordpress.com/2021/10/06/nepals-indigenous-majhi-communities-demand-cancellation-of-sunkoshi-2-hydropower-project/, accessed December 14, 2023.

13 For more information on the project, see the dedicated Nepali Army's webpage, or the project's assessment on EJAtlas.org.

14 *Indigenous Newar communities in Khokana and Bungamati call on UN mechanisms in Geneva for immediate actions to protect their rights threatened by the construction of Fast Track Expressway and other infrastructure projects.* (December 1, 2020). Community Empowerment and Social Justice Network. https://cemsoj.wordpress.com/2020/12/01/indigenous-newar-communities-in-khokana-and-bungamati-call-on-un-mechanisms-in-geneva-for-immediate-actions-to-protect-their-rights-threatened-by-the-construction-of-fast-track-expressway-and-other-in/, accessed December 14, 2023.

15 Shrestha, S. (May 4, 2018). Our Land Is Us, We Are Our Land. *Nepali Times.*

16 Government of Nepal: Ministry of Physical Infrastructure and Transport. (2015). *Kathmandu-Terai/Madhesh Fast Track Road Project: Environmental Impact Assessment Final Report – Volume I: Main Report.*
17 Kumar Mandal, C. (March 8, 2018). Khokana Residents Protest against Kathmandu-Tarai Expressway. *The Kathmandu Post.* https://kathmandupost.com/national/2018/03/08/khokana-residents-protest-against-kathmandu-tarai-expressway, accessed December 14, 2023; Republica (September 12, 2020). Symbolic protest against Kathmandu-Tarai Fast Track Project at Kokhana. *MyRepublica.* https://myrepublica.nagariknetwork.com/news/photos-symbolic-protest-against-kathmandu-tarai-fast-track-project-at-khokana/, accessed December 14, 2023.
18 *Indigenous Newar Communities in Khokana and Bungamati Call for Follow-up Actions from UN Experts in Geneva to Protect their Rights Threatened by the Construction of Fast Track Expressway and other Infrastructure Projects.* (July 10, 2022). Community Empowerment and Social Justice Network. https://cemsoj.wordpress.com/2022/07/10/indigenous-newar-communities-in-khokana-and-bungamati-call-for-follow-up-actions-from-un-experts-in-geneva-to-protect-their-rights-threatened-by-the-construction-of-fast-track-expressway-and-other-inf/, accessed December 14, 2023.
19 *Chobhar Dry Port.* (n.d.). Community Empowerment and Social Justice Network. https://cemsoj.wordpress.com/human-rights-advocacy/chobhar-dry-port/, accessed December 14, 2023.
20 Tauli-Corpuz, V. (August 2, 2019). Indigenous people are being killed for protecting our land. But we will not be silent. *The Washington Post.* https://www.washingtonpost.com/opinions/2019/08/02/indigenous-people-are-being-killed-protecting-their-land-we-will-not-be-silent/, accessed December 14, 2023, *2021 Global Witness Report: 200 Land and Environmental Defenders Were Murdered in 2021* (September 29, 2022). Business and Human Rights Resource Center. https://www.business-humanrights.org/en/latest-news/2021-global-witness-report-200-land-environmental-defenders-were-murdered-in-2021/, accessed December 14, 2023.
21 *Reflections on Indigenous Participation at COP27.* (November 24, 2022). Land is Life. https://www.landislife.org/reflections-on-indigenous-participation-at-cop27–1326/, accessed December 14, 2023.

References

Dhungana, H., & Maskey, G. (2019). Natural Resources and Social Justice Agenda in Nepal: From Local Experiences and Struggles to Policy Reform. *Public Policy*, 5, 7–31.

Ghale, S. (2018). Men in Green: Conservation Actors Have Overlooked the Army's Excesses in Protected Areas. *Kathmandu Post*, May 22, 2018.

Government of Nepal. (2022). National Foundation for Development of Indigenous Nationalities Act, 2058.

Martinez-Alier, J. (2003). *The Environmentalism of the Poor: A Study of Ecological Conflicts and Valuation.* Northhamption, MA: Edward Elgar Publishing.

McLean, J. (1999). Conservation and the Impact of Relocation on the Tharus of Chitwan, Nepal. *Himalaya, The Journal of the Association for Nepal and Himalayan Studies*, 19(2), 8.

Mishra, H. R. (2004). A Delicate Balance: Tigers, Rhinoceros, Tourists and Park Management vs. The Needs of the Local People in Royal Chitwan National Park, Nepal. In *National Parks, Conservation, and Development: The Role of Protected Areas in Sustaining Society*. Edited by Jeffrey A. McNeely and Kenton Miller, 197–205. Washington, DC: Smithsonian Institution Press.

Müller-Böker, U. (1991). Wild Animals and Poor People: Conflicts between Conservation and Human Needs in Citawan (Nepal). *European Bulletin of Himalayan Research*, 2, 28–31.

Robertson, T. B. (2018). DDT and the Cold War Jungle: American Environmental and Social Engineering in the Rapti Valley of Nepal. *Journal of American History*, 104(4), 904–930.

Sikor, T., Satyal, P., Dhungana, H., & Maskey, G. (2019). Brokering Justice: Global Indigenous Rights and Struggles over Hydropower in Nepal. *Canadian Journal of Development Studies/Revue canadienne d'études du développement*, 40(3), 311–329.

Stevens, S. F. (1993). *Claiming the High Ground Sherpas, Subsistence, and Environmental Change in the Highest Himalaya*. Berkeley: University of California Press.

Thing, S. J. (2019). Politics of Conservation, Moral Ecology and Resistance by the Sonaha Indigenous Minorities of Nepal. In C. Griffin, R. Jones, & I. Robertson (Eds.) *Moral Ecologies: Histories of Conservation, Dispossession and Resistance*, Cham, Switzerland: Palgrave Macmillan, 37–58.

Part 5

Urban Development and Environmental Justice

14 Ensuring Health, Hygiene, and Dignity for Solid Waste Workers

Prashanna Man Pradhan and Bhawana Sharma

Introduction

According to the Constitution of Nepal, solid waste management is a part of public health and sanitation. However, solid waste management has become a major environmental problem for the fast-growing towns and cities in Nepal with many unable to provide satisfactory waste management services.

Modernization, development, and population growth further increase the demand for food, shelter, and other essentials resulting in increasing waste generation. The waste is generally not properly segregated, collected, and treated leading to adverse impacts on human health and the surrounding environment.

At the same time, solid waste management is one of the sources of income for people who segregate and resell waste. In Nepal, as in many developing countries, the informal sector plays an important role in waste management. Hundreds of waste pickers depend on waste for income and survival despite the negative repercussions on their health and dignity. Though resource recovery from such waste is one of the best options for the circular economy, it is practiced informally, haphazardly, and in a way that jeopardizes the health of the waste workers. Sanitation workers play a critical role in maintaining hygiene and managing waste, but they often face hazardous working conditions and social inequalities (Dangi et al., 2006).

Waste picking is entrenched as a caste-based occupation in Nepal. It has been deeply stigmatized as "work for untouchable people." Formally, the government of Nepal legally abolished and criminalized any caste-based discrimination including "untouchability" (the ostracism of a specific caste) in 1963 (Aahuti, 2010). This ban was reaffirmed under the Untouchability (Offence and Punishment) Act of 2011 and in the 2015 Constitution. Despite this legal prohibition, informal caste discrimination continues to exert a major toll. Discrimination based on Dalits' perceived untouchable status is still rampant in the informal sector in rural areas. Together with economic and ecological crises in the hinterlands, many Dalits flee for the cities to escape caste discrimination (Maharjan and Amrit Man, 2019).

Migrant workers, homeless people, and other marginalized groups make up much of urban waste workers. These belong to many regions, ethnicity, caste, and social or ethnic background but lack organizations to protect their health and

DOI: 10.4324/9781003371175-19
This chapter has been made available under a CC-BY-NC-ND 4.0 license.

well-being. Therefore, they suffer social discrimination, insecurity, and a lack of legal protections.

Because they do not have citizenship cards, many waste workers lack access to – or even often information about -- relevant government services such as unemployment insurance, disability benefits, and pensions. Furthermore, only workers employed in the public sector are covered by the public pension system, and this group comprises only 4% of the population (Sapkota et al., 2020).

This chapter illustrates the plight of informal solid waste workers in Nepal as an example of occupation-based environmental injustice. It also profiles efforts to achieve environmental justice by and on behalf of these workers.

Policy Framework

In October 2018, The World Health Organization (WHO) released the first-ever Guidelines on Sanitation and Health. The guidelines were developed because sanitation programs have not been achieving anticipated health gains and there was a lack of authoritative health-based guidance on sanitation. They set out four principal recommendations: (1) Sanitation interventions should ensure that entire communities have access to sanitary toilets on site. (2) The full sanitation system should undergo local health risk assessments to protect individuals and communities from exposure to human waste – whether this be from unsafe toilets, leaking storage, or inadequate treatment. (3) Sanitation should be integrated into regular local government-led planning and service provision to avert the higher costs associated with retrofitting sanitation and to ensure sustainability. (4) The health sector should invest more and play a coordinating role in sanitation planning to protect public health. Solid waste management directly lies under the broad section of Sanitation and Health but lacks adequate clarity and specificity.

Goal 6 of the United Nation's 17 Sustainable Development Goals (SDGs) calls for "clean water and sanitation for all." Unless and until there is safe management of solid and liquid waste there is always a risk of contamination of groundwater. Waste workers are the essential public service providers involved in waste management and therefore crucial for the achievement of the SDGs. SDG 6 also pronounces "for all" putting forward the idea of no one left behind as the core agenda.

Nepal's Labor Act 2074 and the Labor Rules 2075 also set out a legal framework for protecting the rights of workers, including provisions related to workplace safety and health. Furthermore, the Constitution of Nepal 2015 recognizes healthy environment as a fundamental right. This includes the right to clean air, water, and food, and the right to live in a healthy environment. The Constitution also provides for the right to work in a safe and healthy environment.

The Government of Nepal has issued a set of rules by exercising the power conferred by section 50 of the Solid Waste Management Act, 2068 (As per Nepal's Law Commission), Solid Waste Management Rule 2013. The entire set of rules brings in the segregation and harmful waste management process, including penalties for violations. But the law fails to address adequately the rights of the people who are working directly in the waste management sector. The Health Care Waste Management Guideline of the Ministry of Health and Population includes generalized language about potential health hazard, but social, economic, and long-term health hazards are not detailed (Sapkota et al., 2020).

Thus, Nepal does have a legal framework to protect solid waste and other precarious workers addressing environmental justice in occupational settings but they need to be fully realized. The following case study illustrates both the challenges and possibilities of one intervention to protect the health, well-being, and dignity of marginalized solid waste workers.

Case Study: Solid Waste Work as Environmental Justice Challenge

While some waste pickers in Nepal are hired through private companies, most work informally and individually. There is no exact data on the number of informal waste workers but a few studies have estimated them at between 7,000 and 15,000 (Huynh et al., 2022). These waste workers are engaged in waste collection, separation, waste rickshaw pulling, sweeping, and waste carrying (Rijal et al., 2014). Because they work in the informal sector, they lack employment contracts, job security, and social protection. This informal nature of work further exacerbates the challenges they face in terms of safety and equity (Dangi et al., 2009).

Whether these waste workers are formal or informal, they often work in unsanitary and hazardous environments, exposing them to various health risks. They handle waste, including medical and hazardous, without adequate Personal Protective Equipment (PPE) or proper training, increasing their vulnerability to injuries, infections, and occupational diseases. They often face low wages, limited access to benefits, and job instability.

Sanitation workers also often face social stigma and discrimination due to the nature of their job. This is particularly true for those engaged in manual scavenging, which is still prevalent in some parts of Nepal. Manual scavengers, who manually clean and empty pit latrines and septic tanks, waste pickers, street cleaners, and others, face discrimination and exclusion from society. Many informal and temporary waste workers operate with little to no formal training on the occupational health risks of their work. While there is some resistance to these conditions, many have come to accept them as they lack other options.

Kirtipur is an ancient town in Kathmandu Valley. It is also home to Nepal's biggest public educational institution, Tribhuvan University. Many people choose to live in Kirtipur while working or studying in Kathmandu due to its relatively lower cost of living and its historical and cultural significance. The proximity to Kathmandu and the availability of transportation connections make Kirtipur an attractive residential option for those seeking to avoid the crowd of the capital. As a result, Kirtipur also attracts tourists and visitors who come to explore its cultural treasures. The influx of tourists during peak seasons can contribute to crowding and excess waste generation. It has one of the highest growth rates in the country at 4.47% (Karki et al., 2022). Overcrowding and population growth have resulted in increased waste generation, attracting many informal waste workers.

Figure 14.1 Sanitation worker cleaning himself before the construction of the WASH facility in Kirtipur.

Photo credit: ENPHO.

The Kirtipur Municipality has recently constructed a segregation center for solid waste. The center has a waste shredding machine as well as provision to turn trash into organic fertilizer. They also sell the scraps to wholesalers. There are some 50 waste workers directly involved and many others in various capacities at the waste transfer station. Most of them are from ethnic and low-caste groups with poor socio-economic status and low literacy rates. Many are domestic migrants and therefore lack the protection of local social networks.

Unfortunately, these waste workers were not provided with basic sanitation facilities such as toilets, personal protective equipment, safe drinking water, hand washing, laundry, and bathing. Women face the added challenge of not being able to change clothes regularly or during menstruation. The lack of access to proper sanitation facilities and clean water not only affects their physical health but also their safety and dignity (Figure 14.1). Small children often accompany their mother placing them at risk for injury and health impacts. Due to the demand of the workers, the municipality constructed a public toilet at the facility but it went out of service in the lack of proper management.

To address these challenges, the Environment and Public Health organization (ENPHO) has initiated a Water, Sanitation and Hygiene (WASH) project. ENPHO is a service-oriented scientific and research-based national non-governmental organization working in the field of Environment, Water, Sanitation and Hygiene (WASH), and public health in Nepal since 1990. It promotes integrated community-based approaches for safe water, sustainable sanitation, improved hygiene behavior, clean indoor air, and safe living environment. One of its highlights is demonstrating effective management of fecal sludge, wastewater, and solid waste including promotion of recycling and safe reuse. ENPHO provides technical assistance and demonstrates models that can be replicated in the community. ENPHO works closely with local, provincial, and national levels of government to help meet Nepal's Constitutional goals on health and environment as well as international frameworks such as the UN's Sustainable Development Goals.

Prior to working in a specific municipality, ENPHO creates a memorandum of understanding with relevant government bodies spelling out roles and responsibilities of everyone involved. In the case of the Kirtipur WASH facility, ENPHO developed a MoU with the municipal authority, which then worked with private waste management partners who are beneficiaries and the implementor for the WASH facility.

The site of the new WASH facility used to be a waste collection, segregation, and resell center for Kirtipur Municipality. The Solid Waste Management Center is located on the property of Tribhuvan University and is near a major bus stop, making it accessible to waste workers. The city outsourced management of household-level solid waste to four different private partners who are waste collectors turned traders. These companies were segregating waste and transferring it to vendors or to landfill sites. Additionally, the center also collects biodegradable waste from the town and turns, them into organic fertilizer (compost).

More than 50 sanitation workers provide invaluable services through these private partners. Apart from the regular waste pickers and traders, there are many others visiting the collection center for potential work in waste trading.

During the design of the WASH facility, there were several rounds of discussions with the municipality and private waste management companies to identify needs and opportunities for improved working conditions at the facility. During the handover of the project, ENPHO provided the municipality with the business plan to run the facility. As it was a pilot project with highly sophisticated facilities, the projected construction and maintenance cost was high. Therefore, significant design efforts focused on lowering cost while maintaining working condition quality.

To address the issue of equity, ENPHO, other non-governmental partner organizations, and Kirtipur Municipality have adopted city-wise inclusive sanitation (CWIS) which serves as the framework to attract support from bilateral and multilateral donors and philanthropic foundations. The newly endorsed WASH Act and "Kathmandu Declaration of 20 Key Recommendation for Advancing WASH Sector in Nepal" September 2023 clearly spells out gender equity and social inclusion in WASH sector to ensure occupational health, safety, and dignity for frontline sanitation workers.

The facility attempts to mitigate the issue of discrimination by providing services that are often provisioned for more elite caste and social groups. The 710 sq. ft. WASH facility center includes separate toilets with urinals for male and bathing units for male and female sanitation workers, handwashing basins, lockers, sitting lounge, reverse osmosis water purifier, a dug-well, 1,000 litre overhead PVC tank and a 5000 litre reserve tank for consistent water supply, and pump facilities. It provides adequate space, separate areas for toilets and showers, separate lockers, resting rooms, PPE storage, sanitary pad provision, and laundry facilities with washing machines. This allows all employees to enjoy a safe and hygienic environment that promotes the well-being of the associated workers (see Figure 14.2).

Providing a WASH facility for waste workers is essential to ensure their safety. To address their health risks due to exposure to hazardous materials, the facility also provides regular health check-ups, access to medical facilities, and necessary vaccinations. When waste workers receive proper healthcare it supports their well-being and helps counter the negative perceptions associated with their occupation. This well-equipped WASH facility allows waste workers to clean themselves thoroughly after work, ensuring that they are protected from harmful bacteria and germs. By offering better facilities such as clean and safe working environments, protective gear, and proper sanitation facilities, the overall working conditions for waste workers have been significantly enhanced. The center can become a model to develop awareness campaigns and educational programs about the importance of waste management and the vital role waste workers play in maintaining cleanliness and hygiene, such that the value of waste workers is realized. Some waste workers think lowly of themselves. One of the waste workers said, "We were hesitant to use such a sophisticated facility at the beginning, we always thought such facilities are for rich and powerful." Establishing mechanisms to acknowledge and appreciate the efforts of waste workers can go a long way in boosting self-esteem.

In interviews, some of the waste workers expressed their appreciation for the new facility.

One worker commented, "We always felt neglected, but this facility shows the value of our existence." A female worker noted, "Changing clothes was a struggle, it was worse when we were menstruating. Now we can change, and sanitary pads are also available." Another worker stated, "After the work our landlords were reluctant to let us in as we were dirty and smelly. Now we can go home clean, we can even wash our clothes here."

Municipality authorities also expressed their gratitude.

Former Mayor Ramesh Maharjan said,

> Solid waste management is one of the indicators for any livable cities. The municipality has commenced many initiatives under solid waste management, and we are happy to run the facility under a sustainable public private partnership model. This can be showcased as the learning center for students studying social sciences.

Subha Laxmi Shakya, the current Deputy Mayor added, "This facility also considered the need of the female workers which is highly appreciable."

Despite facing several challenges such as budgetary issues, the complexities of cooperative management, and the economics of the waste trade, the Kirtipur WASH facility has successfully been transformed from a mere concept to reality. After initial funding by ENPHO and the International Non-Governmental Organization BORDA South Asia, the project was handed over

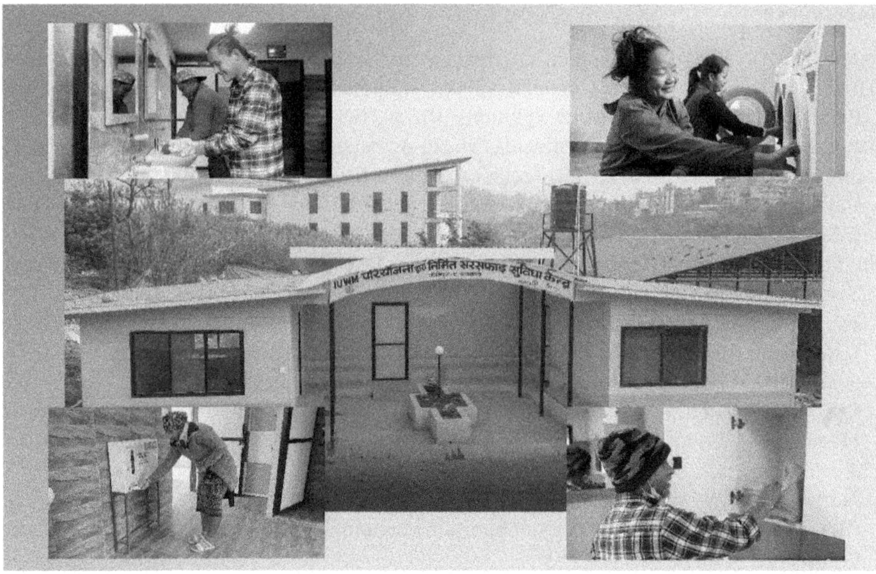

Figure 14.2 The new WASH facility in Kirtipur.
Photo credit: ENPHO.

to the municipality for sustainable management and public funding. To aid in this process, ENPHO provided a comprehensive business plan to the city. The facility is quite expensive to run with an operating cost of about USD $1,500 per year. The plan considered several options: a city-run facility accessible only to the waste workers, a cooperative public-private partnership with access to workers and the general public and a fully outsourced model. The city ultimately decided that a public–private partnership model was the most sustainable such that private waste-carrying vehicles and the segregation center provide tipping fees per the vehicle and, because of its prime location in Kirtipur, many visitors also use the facility with a small users' fee.

Conclusion

A wide range of studies have shown that people feel a sense of psychological security if they have safe and dignified work environments (Marello & Helwege, 2018; Ma & Hipel, 2016). Sanitation workers provide essential public service, yet they remain stigmatized and discriminated against. They are subject to a vicious cycle of poor working conditions, injuries and fatalities, and financial instability. Therefore, it is clear that improved health protection and sanitation facilities are prerequisites to accessing employment benefits and securing dignified working conditions. Such improvements will be necessary for Nepal to meet its aspirations to achieve the SDGs and to fulfill its commitments established in the 2015 Constitution and its associated laws and policies. The Kirtipur WASH facility is one such innovative intervention.

References

Aahuti. (2010). *Varna System and Class Struggle in Nepal.* Kathmandu: Samata Foundation.

Dangi, M., Urynowicz, M., & Cohen, R. (2006). Kathmandu's Solid Waste: Engineering and Policy Analyses for Sustainable Solutions. *Waste Management Research,* 10.

Dangi, M. B., Cohen, R. R., Urynowicz, M. A., & Poudyal, K. N. (2009). Report: Searching for a Way to Sustainability: Technical and Policy Analyses of Solid Waste Issues in Kathmandu. *Waste Management and Research,* 27(3): 295–301. https://doi.org/10.1177/0734242X08094951.

Huynh, T. B., et al. (2022). Association between Informal Employment and Depressive Symptoms in 11 Cities in Latin America. *SSM-population Health,* 18, 101101.

Karki, A., et al. (2022). Mental Health Risks among Informal Waste Workers in Kathmandu Valley, Nepal. *Inquiry,* 59. https://journals.sagepub.com/doi/full/10.1177/00469580221128419

Ma, J., & Hipel, K. W. (2016). Exploring Social Dimensions of Municipal Solid Waste Management around the Globe: A Systematic Literature Review. *Waste Management,* 56, 3–12.

Maharjan, J., & Amrit Man, N. (2019). Sustainable Energy Planning at Municipal Level: A case study of Residential sector at Kirtipur Municipality. *Proceedings of IOE Graduate Conference* 6. ISSN: 2350–8914.

Marello, M., & Helwege, A. (2018). Solid Waste Management and Social Inclusion of Wastepickers: Opportunities and Challenges. *Latin American Perspectives,* 45(1), 108–129.

Rijal, C., Kishor, A., & Bhattarai, N. K. (2014). *A Study on the Collection of Waste PET Bottles in the Kathmandu Valley.* Kathmandu: Himalayan Climate Initiative.

Sapkota, S., Lee, A., Karki, J., Makai, P., Adhikari, S., Chaudhuri, N., & Fossier-Heckmann, A. (2020). Risks and Risk Mitigation in Waste-Work: A Qualitative Study of Informal Waste Workers in Nepal. *Public Health in Practice,* 1, 100028.

15 Urban Environmental Justice

For Whom, from Whom?

Kirti Kusum Joshi

Introduction

Nepal is experiencing rapid urbanization, with massive conversion of rural land for urban uses as a prominent feature. When urban growth sprawls over nearby hinterlands and consumes natural resources in the process, many would call it 'development' – 'haphazard but still development'– but only a few would understand its threat to people's livelihood. Besides sprawl, the urban exploitation of rural areas comes in many different forms and manifestations – it could be industries such as brick kilns that deplete and pollute natural resources at city outskirts to produce goods for urban consumption, or it could be water supply or waste disposal projects in rural hinterlands that serve nearby urban populations.

Economics often guide urban policymaking. The primary role of an economist is to analyze facts and supply analysis to policymakers for necessary actions. The normative field of welfare economics aims at evaluating social desirability of alternative social situations by examining whether the allocation of resources maximizes social welfare or not. Economists often rely upon Cost Benefit Analysis (CBA) for this task. Contesting the dominance of economics in policymaking is within the "Environmental Justice" (EJ) framework which critiques economism for an inadequate analysis of structural factors that favors the economy toward capitalist elites and against the subaltern classes.

Despite a long history of the global EJ movement that dates back to the late 1970s (Fragkou, 2019), EJ as a movement is still new in Nepal (See Adhikari, this volume). For many years, the lack of civil rights in Nepal, such as free press, right to information, freedom of expression, and right to participate in an organization or group had prevented marginalized peri-urban communities from protesting strongly against the urban invaders (Ghimire, 2003). The promulgation of the 2015 Constitution has created two major breakthroughs – first, inclusion of 'right to clean environment' as a citizen's fundamental constitutional right, and second, empowerment of local governments as 'the people's government'. However, even today, many cases of environmental injustice remain unresolved, with little or no interest among policymakers and local governments (see Mainaly, this volume).

Can innovations in economic analysis in some way bridge the gap between 'business as usual' policymaking and environmental justice advocacy in Nepal,

DOI: 10.4324/9781003371175-20
This chapter has been made available under a CC-BY-NC-ND 4.0 license.

and if so, how? This chapter seeks to answer this question by discussing the scope and limitations of cost-benefit analysis to capture the environmental costs facing Nepal's most disadvantaged groups. Such discourse is essential because in Nepal, project feasibility is often determined based on economic analyses, and this applies to environmentally sensitive projects as well.

To examine these important questions, this chapter examines two cases of urban environmental injustice. The first case deals with the abandoned Sisdol (also spelled as Sisdole) landfill site where Kathmandu Valley municipalities disposed of their waste for 17 years despite promising to use the site for only three years. The Sisdol residents have long been demanding compensation for their sufferings including degraded quality of life and exposure to chemicals linked to cancer and serious skin diseases, among others. However, ignoring local demands to clean the site and compensate them for the damages, the state authorities proceeded with the plan to use the newly constructed Bancharedanda (also spelled as Banchare Danda) landfill site, which is located only about 2 km away from the old site. From an environmental justice viewpoint, the Sisdol case shows how poverty makes a community vulnerable to exploitation even by the state. From an economic perspective, the Sisdol case suggests failure in terms of valuating project gains and losses from social welfare perspective in addition to not considering project uncertainties.

The second case looks into the conflict between residents and brick kilns in parts of Kavre district. Most brick kilns here operate in close proximity to human settlements. For years, locals have been demanding a ban on brick kilns, citing adverse effects on environment, public health, agricultural productivity, and animal health. Many residents including disadvantaged groups are small landholder farmers. The brick kiln operators, on the other hand, claim that they are benefitting society by supplying affordable bricks and employing poor migrants. Moreover, local governments also stand to earn revenues from the factories. The question is whether the gains outweigh losses and for whom, given that a basic CBA tends to overestimate economic gains and undervalue social and environmental losses, especially in regard to the distribution based on class, class, ethnicity, and geographic location.

Welfare Economics and Environmental Justice

With increasing income inequality and concentrated poverty, prioritizing justice over mere economic efficiency in evaluating urban policies is crucial (Steil and Connolly, 2019). The common framework of EJ incorporates several interrelated and mutually supportive dimensions, including distributive, procedural, and restorative justice, among others (Schlosberg, 2007).

Distributive justice refers to the fair and equitable distribution of environmental benefits and burdens among different groups of people. Its aim is to prevent or reduce the disproportionate exposure of some groups, such as low-income, minority, or indigenous communities, to environmental hazards and risks. In many instances of CBA, there are concerns that a singular focus on reducing absolute costs would lead to redistributing risks equally within social groups without addressing the underlying system that generates these risks (Fragkou, 2019).

Procedural justice refers to the fair and inclusive participation of all stakeholders in decision-making processes that impact their environment and health. Participatory justice, on the other hand, entails giving affected communities a decisive voice in policies and practices affecting their environment and ensuring that their input is considered in decision-making processes. Procedural and participatory justice are important from an economic perspective because participation of all stakeholders including subaltern classes helps in understanding the real costs and benefits of a decision and its distributional effects, and also helps prevent costly conflicts.

Environmental justice is not just an approach to avoid bias among people of different backgrounds but a proactive way to rectify historical and structural wrongs. This involves identifying the forces that produce these inequities and developing new systems to address them. A basic CBA tends to overlook structural inequalities that market dynamics have generated over time and consequently falls short in addressing these historical and pre-existing inequalities. This is where the concepts of corrective, restorative, and transformative justice come into play.

Corrective justice seeks to punish offenders for their wrongful actions, in order to restore balance and order to society. Restorative justice, on the other hand, refers to the repair and restoration of the harm that past or present environmental injustices have done to people and nature. Transformative justice aims to challenge and transform the dominant paradigms, institutions, policies, and practices that generate and reproduce environmental inequalities and conflicts.

In its basic form, CBA aims to capture costs and benefits of a decision at the aggregate level without looking into the differential impacts of the decision to different groups in the affected community. Assigning different weightages to costs is, therefore, necessary to capture disproportional impact on disadvantaged groups as a measure to rectify historical wrongs committed against them. An economic change is efficient if the winners from a policy change could hypothetically compensate the losers and still be better off – it is important, then, not to overlook any winners or losers, while also ensuring that compensation actually takes place and is adequate (Hanley, 1992). This is where the concept of externalities comes into play. Externalities are gains or losses that people or society experience as a result of an activity but which are missing in the activity's market price. Most of the environmental problems, such as pollution, noise, and traffic congestion, are examples of negative externalities: they exist because the market fails to capture the true costs of an activity, resulting in overproduction or overconsumption of goods and services that harm the environment and public health.

Economists often suggest the state to impose tax or other regulations to internalize such externalities in alignment with the so-called 'polluter pays' policy. Moreover, at times, market biases can result in harm to the most disadvantaged populations, warranting the consideration of public subsidies. As we will see in the following cases the willingness and capacity of the state to provide for these protections is often in question.

Case 1: Urban Waste Disposal

On June 5, 2005, which was ironically the World Environment Day, Kathmandu Metropolitan City (KMC) began disposing Kathmandu Valley municipal waste at Sisdol landfill site in Nuwakot's Okharpauwa for a planned duration of three years. However, waste disposal continued for 17 years, overwhelming the site with hazardous trash and causing local residents to suffer severe health problems including headaches, diarrhea, respiratory problems, skin diseases, and even cancer. The landfill site's leachate also reduced agricultural production and led to diseases in livestock. In 2019, the government started building a new landfill site nearby in Bancharedanda, abandoning the original site as a malodorous mound and the local residents as if they were lesser citizens. As a local resident shared in a media interview, "Once a paradise, Sisdol is now a hell. We have nowhere to go" (Annapurna Post, 2017, November 2). Another resident had this to say: "The state doesn't consider us humans" (Awasthi, 2022, February 22).

The Sisdol area residents belong mostly to Tamang, Balami, and Hill Brahman ethnicities – Tamang and Balami (which is a subset of Newa community with a distinct language) are both minority ethnic groups. While there is no explicit evidence that the presence of minority ethnic groups directly caused the state authorities to exploit Sisdol for years, it is plausible that under-development and ethnicity-based discrimination contributed to such exploitation. Most people in the area are small farmers. The literacy rate is below one-third. Road accessibility and connectivity is poor. These factors suggest a class dimension to the crisis.

At first, it was the local residents themselves who responded positively to the KMC's call for a new landfill site, hoping that this would open the door for Okharpauwa's development. The KMC's proposal was to use Sisdol site for three years only: the Bancharedanda landfill site had to be ready by then. The head of Okharpauwa Sanitary Landfill Site Coordination Committee recalled in a media interview, "Our joy knew no bounds. We even greeted garbage trucks with garlands. We now feel remorseful about it" (Awasthi, 2022, February 22).

The government and KMC had made a series of promises to the affected households in Sisdol, including constructing a hospital, enhancing roads and other physical infrastructure, and providing jobs for each affected household (Ojha, 2022, May 31). KMC also separately agreed to adopt scientific measures to prevent leachate from polluting the river and to control foul odors. Likewise, the government had announced land acquisition plans in the impacted area for relocation of the affected households. With these promises not yet realized, KMC began disposing of waste at the Bancharedanda site in 2022, with up to 350 garbage trucks arriving at the site every day. The new site is merely 2 km from Sisdol. This has frustrated local residents, leading them to engage in frequent protests, disruption of garbage collection, and even vandalization of garbage trucks, resulting in altercations with security forces.

Sisdol is situated about 23 km north-west of KMC. While not very far in physical space, it is politically distant enough for the Kathmandu Valley's policymakers, mayors, and residents to overlook the inhumane conditions there. While the state authorities have kept blaming one another (Kunwar, 2019, September 23), the residents, along with their political representatives, have found themselves helpless. Even KMC's written pledge regarding the closure of the Sisdol site has yielded no action (Tamang, 2023, March 3). This implies the institutionalization of discrimination in Nepal against the poor and politically disenfranchised. Otherwise, how could the state authorities disregard the Sisdol residents' right to clean environment? Even in terms of media coverage, Sisdol gathers attention only when the residents obstruct garbage trucks. This indicates that Sisdol's issue has not yet gained broader and sustained attention.

The Sisdol residents' demands are reasonable and in line with their constitutional rights: all they want is the old site's closure, health insurance for people and cattle, and relocation of seriously affected households (Kunwar, 2019, September 23). These demands and the KMC's own promises could have been a part of corrective and restorative justice. As the country's richest local government and the largest waste producer, KMC should have taken proactive steps to address the Sisdol issues. Moreover, it would have made an economic sense for KMC to do so.

Waste disposal facilities at Sisdol and now Bancharedanda have provided relief to the Kathmandu Valley's municipalities and its residents, but people living near the landfill site are suffering. They gain nothing from the dumps but bear many costs. From an economic perspective, waste disposal is an example of negative externalities: although households in the Kathmandu Valley pay waste management fee to service providers and the latter pay taxes to the state authorities, the pricing mechanism does not include external costs that the residents near landfill sites are bearing in the form of health problems and degraded environment. Such imbalance between costs for the locals and benefits for the Kathmandu Valley's residents implies a violation of distributive justice, creating a situation of unfairness.

The state authorities have chosen to refrain from directly engaging local residents, fearing that the latter's demands will only increase if they come to the negotiating table. However, participation of the affected groups in the decision-making process is not only important from the EJ perspective but it also makes economic sense for the authorities themselves. Such participation helps reduce the risk of future conflicts, and thereby save time and effort in managing waste.

Corrective and restorative measures such as compensation and remediation of the old landfill site require huge investments, but there are ways to generate funds. Most households in the Kathmandu Valley pay nominal monthly fees to waste collectors at about Rs 200 (US$1.50) per household, but gain a benefit that far outweighs the costs borne by the residents near the landfill sites. Given that the household waste collection has already become a lucrative business in the Valley, state authorities have an opportunity to find financial means to fund corrective and restorative measures. Would the Valley households be ready to pay more for waste collection? As per the 'polluters pay policy', they should.

Substituting a fixed monthly fee, which is in practice now, for a volume-based fee would assist in reducing waste at its source, and also promote recycling and composting. This will provide funds for restorative justice to support building healthy, prosperous, and equitable communities in places once considered 'sacrifice zones' (Faber, 2008). More broadly, it could catalyze transformative justice that aligns with the overarching objective of rendering a produce-consume-dispose treadmill of destruction cycle obsolete: this will make large landfill sites unnecessary, thereby shifting from the 'not in my back yard' (NIMBY) perspective to a 'not in anyone's backyard' approach. Reducing waste and toxicity at the source will aid in mitigating health hazards and subsequently lower environmental costs near waste disposal sites. Additionally, this effort will play a role in prolonging the operational lifespan of waste management facilities. This extension can serve to rationalize the initial substantial investments and offer an extended timeframe for the accrual of benefits and co-benefits resulting from any government investments in local infrastructure, health, and education in the affected area.

Case 2: Brick Kilns: Smoking at the City Outskirts

Just outside the Kathmandu Valley, Kavre residents, particularly in Panchkhal and Namobuddha municipalities, have been dealing with severe air pollution including toxic gases and black carbon particles from nearby brick kilns. The residents have long been demanding a ban on these kilns, citing adverse effects on the environment, public health, agricultural productivity, and animal health. Moreover, the brick kiln operators also face accusation of contravening the legal rule that requires them to operate at least 500 meters away from a human settlement.

Brick kilns have been contributing to the local economy by paying taxes to the municipal authorities, which often makes the latter hesitant to oppose them. In the face of the municipal authorities' indifference, residents have sometimes taken matters into their own hands by dismantling brick kilns. In response, Panchkhal has now formulated legal provisions for compliance with environmental standards, and Namobuddha has altogether banned the operation of brick kilns. But the brick kiln operators have not entirely halted their operations (Figure 15.1).

In a media interview, the leader of brick kiln operators in Kavre noted that in the early 2000s, local residents "were content because they could buy bricks at a reasonable price, and some even made greater profit leasing land to brick factories instead of farming" (Nawasanket Online, 2018, June 8). According to him, the locals started raising the issue of environmental degradation only after exhausting all the benefits they had gained from the factories. The brick kiln operators also claim that they have been providing great service to the broader society by providing jobs to unskilled migrants.

From an economic perspective, employment, tax payments, and availability of cheaper bricks are benefits of brick kilns. Some landowners also benefit from land rent. On the other hand, the costs include health hazards and reduced agricultural productivity because of severe air pollution including emissions of dust and smoke particles. Particulate matter has detrimental effects on human health, causing a variety of respiratory complications as well as cardiovascular and neurological problems: children and pregnant women are particularly vulnerable to such emissions (Eil et al., 2020). Given these health risks, even if some residents might

Figure 15.1 Brick kilns polluting the air in the Kathmandu Valley.
Source: Nepali Times.

have benefited from the brick factories at some point, their economic 'gain' is more likely to be outweighed by their health-related losses over years.

In addition to health costs, small landholder farmers in the area have been experiencing economic losses. According to media reports, due to dust pollution, fruit plants have ceased to flower (Lama, 2018, June 26), and it is no longer feasible to grow leafy vegetables and potatoes (Ghimire, 2018, June 1). In Panchkhal, the areas most affected by brick kiln emissions are where small farmers of ethnic minority backgrounds live: primarily Sarki people (a Dalit caste) and Dunawar and Tamang people (indigenous ethnic groups).

Environmental justice advocates and environmentalists tend to prefer the direct route of "stopping polluters." Some economists, on the other hand, tend to believe that certain amounts of negative externalities (pollution from brick kilns in this case) could be necessary for the affordability and availability of bricks, which are an integral construction material for building houses. This does not mean that economists defend pollution or injustice. On the contrary, the objective of an economic policy is to internalize or reduce negative externalities either through a pricing mechanism or through improved regulations and technologies. The problem is: who benefits and who pays the cost?

For distributive justice, the residents deserve compensation for pollution-induced health and economic losses just as much as the brick kiln operators deserve penalties for generating pollution. It is also important to check that the prevailing environmental standards are scientifically valid, otherwise there is no point blaming kiln operators alone for pollution. The local governments should develop a mechanism, such as insurance package, to compensate farmers for poor harvests, ensuring that those who are most disproportionality affected receive a greater share of compensation.

For legally running brick kilns, state authorities should either incentivize kiln operators to adopt pollution reduction measures or help them shut down their operations by providing alternative technologies. From the cost-benefit analysis, this would ensure citizens' rights to a clean environment ('benefit') while also compensating the industries ('cost') for their investments, thereby contributing toward restorative justice.

It is also important to recall that prevention is better than cure. It is not only the industries that should maintain a safe distance from settlements, but it is also equally important to ensure that residential development does not expand into the existing industrial territory, necessitating effective zoning and land use laws. The new Constitution has empowered local governments to implement zoning, which is an economic tool to separate residential and industrial land uses for reducing negative externalities arising from pollution. It would, however, be challenging for local governments to ensure that pressure groups, including landowners, brokers, and industries, do not influence zoning policies.

Local governments should also be mindful of real estate speculators who might be purchasing inexpensive land near brick kilns with the intention of displacing the factories in the future, once there is enough residential development to build opposition. Finally, local governments should also address the issue of informal settlement growth near brick kilns as poor migrants and landless people are vulnerable to exploitation by industries.

Conclusion

Environmental policymaking in Nepal, as elsewhere, involves a complex interplay of economic, ethical, and political factors. Economics often plays a guiding role, with CBA being a popular and almost a default tool. However, CBAs, in their basic form, often overemphasize efficiency and consider that an economic change is efficient if the total economic surplus (or benefit) of the change is positive, even if some individuals are worse off as a result. It sometimes treats environmental costs as externalities. This goes against the spirit of environmental justice.

Despite its limitations, CBA can be an important tool for identifying and weighing the effects of a project, providing valuable information for policymakers. This is especially true if including the valuation of environmental damages, which should be an intrinsic aspect of CBA, prevents them from being overlooked or deemed insignificant. Therefore, it is worth considering how modifications in CBA to reduce environmental externalities can bridge the gap between economic and environmental justice perspectives.

Possible best practices for a reformed CBA include assigning disproportionately high weightage to the economic losses of disadvantaged groups for distributive justice, internalizing negative externalities concerning outsider beneficiaries for procedural and corrective justice, recognizing participation as a cost-saving measure in the case of participatory justice, and investing a share of future benefits for restorative and transformative justice.

References

Annapurna Post (2017, November 2). Narka banyo Sisdol [Sisdol turned hell]. *Annapurna Post*. Retrieved 19 August, 2023 from https://annapurnapost.com/story/83233. (in Nepali)

Awasthi, M. (2022, February 22). Mahina Naghda samma pani fohor uthena, vyavasthapanma sadhai kichalo [No Garbage Collection for a Month, Management Crisis Persists]. *Nepal Views*. Retrieved 19 August, 2023 from https://www.nepalviews.com/2022/02/22/42298. (in Nepali)

Eil, A., Li, J., Baral, P., & Saikawa, E. (2020). *Dirty Stacks, High Stakes: An Overview of the Brick Sector in South Asia*. World Bank.

Faber, D. (2008). *Capitalizing on Environmental Injustice: The Polluter-Industrial Complex in the Age of Globalization*. Rowman & Littlefield Publishers.

Fragkou, M. C. (2019). Environmental Justice. In A. Orum (Ed.), *The Wiley Blackwell Encyclopedia of Urban and Regional Studies*. John Wiley & Sons Ltd. https://search.credoreference.com/articles/Qm9va0FydGljbGU6NTEyNzE=?aid=104013

Ghimire, S. (2003). Concept of Environmental Justice in Nepal: Environmentalism of the Poor for Sustainable Livelihood. *Himalayan Journal of Sciences*, 1(1), 47–50.

Ghimire, S. H. (2018, June 1). Bhattako dhuwale dhakyo Panchkhal [Kiln dust blankets Panchkhal]. *Nagarik Dainik*. Retrieved 19 August, 2023 from https://nagariknews.nagariknetwork.com/others/150374-1527815700.html. (in Nepali)

Hanley, N. (1992). Are There Environmental Limits to Cost-Benefit Analysis? *Environmental and Resource Economics*, 2, 33–59.

Kunwar. (2019, September 23). Chaudha barshadekhi vyavasthapan bhayena Sisdolko fohor [Sisdol's Waste Goes Unmanaged for 14 Years]. *Naya Patrika Daily*. Retrieved 19 August, 2023 from https://nayapatrikadaily.com/news-details/25822/2019-09-23. (in Nepali)

Lama, P. (2018, June 26). Ita bhattaka karan pradushan badhaeko bhandai Panchkhalvasi andolit [Panchkhal locals protest against brick kilns for pollution rise]. *Lokpath*. Retrieved 19 August, 2023 from https://www.lokpath.com/story/50478. (in Nepali)

Nawasanket Online (2018, June 8). Interview with Jitendra Khayamali. *Nawasanket Online*. Retrieved 1 October, 2023 from https://www.nawasanket.com/antarwarta/2018/06/05/17893. (in Nepali)

Ojha, A. (2022, May 31). Sisdole Locals Say They Won't Allow Garbage at Banchare Danda until their Demands Are Met. *The Kathmandu Post*. Retrieved 19 August, 2023 from https://kathmandupost.com/national/2022/05/31/Sisdole-locals-say-they-won-t-allow-garbage-at-banchare-danda-until-their-demands-are-met.

Schlosberg, D. (2007). *Defining Environmental Justice: Theories, Movements, and Nature*. Oxford University Press.

Steil, J., & Connolly, J. (2019). Just City. In A. Orum (Ed.), *The Wiley Blackwell Encyclopedia of Urban and Regional Studies*. John Wiley & Sons Ltd. https://search.credoreference.com/articles/Qm9va0FydGljbGU6NTEyNzE=?aid=104013

Tamang, S. (2023, March 3). Sisdolma Post-Closing Kaile? [When's the Post-Closing in Sisdol?] *Naya Patrika Daily*. Retrieved 19 August, 2023 from https://www.nayapatrikadaily.com/news-details/108809/2023-03-03. (in Nepali)

16 Cycling for Livelihood in Nepal

Seeking Justice on Two Wheels

Tara Lal Shrestha and Bidhya Shrestha

Introduction

Rapid urbanization is a recent phenomenon in Nepal. In 2021, about 66% of Nepal's population lived in government-designated urban areas (NSO, 2023). However, urban planning is not paying due attention to transportation, especially the dimensions of equity and justice in the mobility of low-income residents. Instead, transportation planning in Kathmandu and many other major cities generally caters to the elite class through building and widening of roads specifically for cars and motorcycles. Subaltern people's need for better facilities for bicycling, which they use to carry out their daily livelihood, is largely ignored. This has prevented the adoption of active transportation infrastructure to support bicycling that is suitable and affordable for those pursuing livelihood activities.

Active transportation such as bicycling has been promoted in both developed and developing countries for reducing emissions from burning fossil fuel and for health benefits (Wood and Roelich, 2019; Lyons, 2020; Adetoyinbo et al., 2022). However, there are still questions about the provision of adequate and equitable infrastructure for this purpose. In developing countries like Nepal, policymakers and planners race to catch up with cities in developed countries by supporting the car culture of the elite class. This has created air and noise pollution and reduced the capacity of cities to support the daily livelihood activities of the subaltern cyclists and other residents.

Studies on active and equitable transportation in cities in developing countries are scarce (Dadashova et al., 2023). To help fill this gap, this chapter shows that people who depend on bicycling for livelihoods face injustice in three ways: exposure to the health risks from air pollution; accidents caused by cars and motorcycles; and, lack of equity in the allocation of resources for the infrastructure (e.g., bike lanes and paths).

As this study deals with inequities in achieving and protecting livelihoods, it can be placed within the framework of 'environmentalism of the poor' (Martinez-Allier, 2003) in an urban context. This concept has been mainly applied in rural context in struggles over the livelihood of the natural resource-dependent people. But this study focuses on the concerns, voices, and activism of these people in improving their condition and in putting pressure on the urban planning system for improvement of infrastructure they require. While the impact of such activism has been minimal so far, these struggles demonstrate the growing agency of these populations.

DOI: 10.4324/9781003371175-21
This chapter has been made available under a CC-BY-NC-ND 4.0 license.

The study identified four types of riders whose members depend on bicycles for their livelihoods: (i) those who use bicycles for going to the office/workplace, (ii) those who use bicycle for door-to-door hawking, (iii) those who use bicycles to deliver food/goods, and (iv) those who, known as *kawadiwala*, collect reusable waste items. These people are poor and belong mostly to marginalized, displaced, or recent migrant populations from the villages of Nepal and India.

The study uses qualitative research methods to examine the livelihood-driven cyclists, their concerns in terms of risks to which they are exposed, and their agency to influence the transportation planning process in Kathmandu. It draws significantly on the experience of the first author who bicycled across Nepal conducting meetings with diverse cyclists and cycling organization members to support advocacy for Nepal's National Bicycle Act. We also conducted eight focus group discussions and 16 key informant interviews with a diversity of cyclists in the above four categories.

Growing Dominance of Car Culture in Kathmandu

Until 1953, in Kathmandu, porters carried bicycles and cars from Bhimphedi crossing Chandragiri hills (Manandhar, 2013; p. 178; Dahal, 2023; Sindhu, 2023). The very first cars were for the king and the ruling class.

Starting in the early 1960s, bicycle culture bloomed in Kathmandu and roads were mainly used for walking and cycling. Government offices, banks, schools, and colleges offered dedicated bicycle parking spaces. Most households had at least one bicycle (Gautam, 2012). With its gentle inclines, beautiful scenery, and convenient distances between key locations across the Valley, Kathmandu has been an ideal cycling city for both commerce and recreation (Subedi, 2023).

Bicycle culture started to change and gave way to car culture starting in the 1980s (Bhattarai, 2018). Now motor vehicles dominate congested roads in Kathmandu making cycling unsafe (Rai, 2011). As such, the bicycling culture has declined significantly and this form of two wheelers has largely become a vehicle for a poor person's mobility (Khanal, 2021). Since the 1990s, the number of motor vehicles has grown at the rate of 12%–13% annually. In 1991, 13.1% of households owned a car; this increased to 30.2% in 2011, whereas cycling declined from 6.6 to 1.5% in that period. In 2021 among 542,892 households in the Kathmandu Valley, 55,271 households (10.2%) had a car, jeep, or van; 263,069 households (48.6%) had a motorcycle/scooter while only 62,494 households (11.5%) had a bicycle.

With the rise of a middle class, there has been a tremendous increase in the number of motorcycles in Kathmandu. According to latest updated data of the Department of Transport Management, out of 3.22 million vehicles registered in Nepal, 78% (2.53 million) are motorcycles and nearly 40% of these ply on the streets of Kathmandu Valley (Wagle, 2021). Because of their fast movements across lanes, motorcycles are prone to accidents and also cause accidents for cyclists.

Dr. Prahlad Yonzon, a renowned conservationist, was killed in October 2011 at Balkhu Ring Road while cycling. Cycling advocates submitted a petition in April 2012 gifting a bicycle to Prime Minister Dr. Baburam Bhattarai and requested him to enact a Bicycle Act. Despite this public support, no such Act was passed and road safety for the cyclists has become even more precarious since. In 2021–2022, 140 cyclists lost their lives and 1,174 were severely injured in road accidents: most of these occurred in urbanized Kathmandu. (Sharma, 2023). In the absence of adequate bicycle infrastructure, regulations, and policies, cyclists face growing risks to their lives (Acharya, 2022; Shrestha & Shrestha, 2022).

Because of a growing pressure from cyclists to address these hazards, Lalitpur Metropolitan City in Kathmandu created shared bicycle lanes. But advocates continue to question the safety of cyclists on the streets of Kathmandu; there are no dedicated bike lanes, and some painted lines in the narrow roadways do not provide sufficient protection (Wagle, 2021).

Cycling and Livelihoods

Many working-class people in Kathmandu use bicycles for their transportation and livelihood activities. In contrast to the latest modern bicycles used by recreational riders, these livelihood-driven cyclists use the older traditional cycles – generally called *budho cycle* (old and cheap).

Cycling for Work: A Diverse Community in Green Mobility

Until the 1980s, both high and low-wage office workers used bicycles to commute. This was because of road safety, relatively lower air pollution and minimal auto traffic. Now, while most middle-class people prefer private motor vehicles, there are still many people of different professions who use bicycles for their daily commuting and livelihood activities. A few high-level job holders also still prefer bicycles due to their physical, mental, and environmental benefits.

Chitra Bahadur Pun, an entrepreneur, said, "I pedal to the office, save at least 50 liters of petrol and reduce at least 100 kg of carbon per month."[1] Jot Narayan Patel pedals 20 kilometers every day from Bhaktapur to Singha Darbar office for fun, exercise and environmental justice. Punam Thulung, a student studying at St. Xavier's College is a dedicated bicycle-lover. After she completed her Bachelor of Management, she began to commute to work on her bicycle. She wishes others would follow her example. "If the road is safer, thousands of girls like me would be on the streets," she said.[2] These types of white-collar cyclists tend to use safety measures (helmet, lock, lights, neon/reflective clothing) and follow traffic signals (Haupt, 2021).

Other lower-income workers use bicycles while commuting to work in Kathmandu to save time and money. They are not aware or, more likely, cannot afford safety gear. Dhani Lal Gupta, a worker in the waste picking and segregating sector, prefers cycling to avoid disturbing passengers on public buses with his dirty work clothes. Cycling also saves him his transportation cost. Mina Tamang, the leader of a group of street cleaners, says, "Even if we take a bath, people can still smell us when we use public bus, so we use active commuting."[3] There are also many regular cyclists from the disabled community because buses, which are their primary means of mobility aren't well-equipped for those with disabilities. Many of these are activist cyclists who are playing leading roles to unite the heterogeneous 'cycling-for-livelihood communities' around the cause of mobility justice, including the building of effective and safe bike paths and lanes and improving traffic safety regulations.

Door-to-Door Vending: Continuing a Tradition of Microenterprise

Door-to-door vending has become one of the key means of livelihood for the urban poor in Kathmandu Valley. This informal sector is an important source of earning livelihood for a large number of poor people, who sell varieties of goods door-to-door. They use bicycles for vending vegetables, fruits, clothes, dry foods, and goods. They pull their fully loaded cycles door-to-door in groups in Kathmandu and ride them home when they finish their vending. They tend to be low-income temporary migrants from Tarai in southern Nepal and India.

Gautam Yadav is one of such hawkers who arrives at the central Kalimati vegetable market every day at 4 am. He does not have time to cook or eat in his room. Instead, he eats in low-cost hotels and rushes to sell vegetables from house to house. Then he begins his second trip from Balkhu wholesale vending fruits door-to-door. Yadav said, "It is not our wish to endure the pollution from dawn to dusk. It is a compulsion." Some hawkers also move around the city with their bicycles asking residents if they need help in repairing pressure cookers and gas stoves and selling pots and utensils. This hawking job requires a relatively low upfront investment for the bicycle and associated equipment (about NRs 2,000–5,000 or $15–38) but earns about NRs 1,100–1,700 ($8.5 to 13) per day (Regmi, 2017: 35). Accordingly, poor people can afford to follow this bicycle-dependent occupation (Figure 16.1).

The first author interviewed more than three dozen hawkers around Kalimati and Balkhu with the help of the second author. Almost all of hawkers consulted for the study said that they have no access to public toilet, proper resting place, and

Figure 16.1 A peddler's bicycle loaded with household goods.

Photo credit: Vijay Ratna Shakya.

drinking water facilities. They have to go to hotels to buy water or carry water from their rooms. Some of them said that most hotels do not allow them to use the toilet. Nirwal Yadav and Ashok Yadav, originally from India who have been hawking in Kathmandu, said "this city is getting more crowded and polluted but we must push bicycle for living as there is nobody to help us."[4,5]

These hawkers face a range of risks, including unsafe roads and city police who prohibit them from selling goods in many public places. Recently, under the directive of Mayor Balen Shah, the Kathmandu Municipality police have confiscated hundreds of bicycles from street vendors. Mass protest did not stop the decision of the municipality which has adopted the policy such that bicycles cannot move on motor vehicle roads and yet has not constructed dedicated bike lanes (Pharak Nepal, 2022). The municipality equates getting rid of hawkers and street food vendors with making the city "clean." The municipality's decision has drawn a lot of criticism but they have made little to no impact (Ghimire, 2023).

Delivery-Cyclists: Adding New Green Mobility Stories in Kathmandu

Gyan Shrestha from eastern Nepal has sold potato chips with the help of *budho cycles* to different shops, in *choks* (cross-roads) and alleys for years. However, pedaling for food delivery is a new phenomenon in Kathmandu. During and after the pandemic, many college students and jobless youths used low-cost bicycles for delivery in the gig economy. As the fuel price rose, many motorcyclists involved in food delivery switched to bicycling. Hundreds of cyclists have appeared on the streets in Kathmandu for food delivery in the past three years. Firms such as Foodmandu and Pathao hired dozens of cyclists for short-distance food delivery. There are about 600 cyclists registered in Pathao for food delivery (Acharya, 2023). Bhojdeals has recently hired 40 cyclists for food delivery. While most cycle delivery workers are male, Pathao has also hired a few female cyclists.

One of the bicyclists from Pathao, Sunil Lama, used to have a motor scooter. He used to earn $8–10 USD per day, which would be exhausted by the daily fuel cost. He sold his scooter and bought an e-bicycle. "Now, I save all money that I earn," he said.[6] After switching to an e-bicycle, he was receiving more than ten orders and earning almost double than before. While the income is good, these delivery staff face peak hour pressure and must navigate in polluted, congested, and unsafe streets. Suraj Lama observes that carrying a big box bag on his back and pedaling in Kathmandu is not easy. He said, "streets are either in disrepair or congested and are full of dirt and dust. The motor vehicles on the other hand speed by recklessly." Indeed, a young delivery Pathaon cyclist Bhim Bahadur Tamang was killed in a road accident in August 2023.[7]

The gig economy has bloomed in Kathmandu and cyclists have revolutionized the takeaway delivery occupation but without assuring safety or security for its riders. They have taken this occupation as an economic necessity – rather than choice – to survive in Kathmandu.

Kawadiwala: The Hidden Recycling Heroes

Approximately 800,000 kilos of waste are produced every day in Kathmandu Valley. Despite a large number of waste recycling firms, tons of recyclable waste are disposed of in landfills. The waste collection processing, and resale work of the *kawadi* is therefore a major source of livelihood for vulnerable populations in Kathmandu Valley. There are more than 700 *kawadi* goods collection centers and 10,000 *kawadiwalas* who collect and resell reusable or recyclable materials in Kathmandu (Nepal Live, 2018). Most *kawadiwalas* rely on bicycles to collect and transport waste.

In the words of Janu Dangol, the representative of *Sar Saphai Jagaran* (a waste management union), *kawadiwalas* work in solid waste landfill sites as informal waste collectors, segregators, *pheriyas* (scrap buyers), cleaners, and in related activities. Despite having low levels of formal education and coming from disadvantaged ethnic and caste backgrounds, they provide essential services for society without any safety or security. Rather than honoring them, they are stereotyped with the term *khates*[8] (derogatory word for homeless ragpickers). In the words of Shanti Tamang, the Chairperson of Sarsaphai Jagaran, "Everyone wants their surroundings clean but they still disrespect *kawadiwalas* and cleaners" (SASAJA, 2022). The *kawadiwalas* vocalize this trauma whenever they get a chance to speak. Sanjaya Shah, a representative of *kawadiwalas*, spoke from the stage on a truck in King's Way Kathmandu, celebrating World Bicycle Day 2022. "We are tirelessly cleaning your city, your discarded dirt and dust, and make your city shine but you call us *khate*!," he said. In front of the mayors of the metropolitan cities of Kathmandu Valley, Shah further said, "We *kawadiwalas,* recycling heroes, manage 30% of the waste of Kathmandu" (Tamang, 2022). The *kawadwalas* roam the city on their bicycles collecting recyclable waste but they are not recognized as essential workers (KhaliSisi, 2023). They pedal for livelihood, not for lifestyle (Serkhel, 2022). They call their bicycle 'the best motor of *majdurs* (laborers)' and continue to clean Kathmandu day after day.

Sukhal Paswan, 40, a Dalit from Bara district in the Tarai, for instance, lives with his five family members in Kathmandu. Despite losing his right hand 25 years ago, he is the primary provider for his five-member family. Being a migrant Dalit in a Kathmandu metropolis, he is not able to make a 'disability card', which would have allowed him to get a social security.

Despite their critical role in the management of the city's waste, the problems of *Kawadi* and their reliance on *budho cycles* are not addressed or protected by local or regional policymakers and planners.[9]

Green Mobility Movement: Act and Activism

Mobile street vendors contribute in critical ways to a sustainable urban supply chain management. The number of such vendors in Kathmandu is large – about 10,000 (Ojha, 2020). The number of buyers of goods from these vendors is also large. In such context, Pitamber Sharma, a planning expert, said "these vendors, who use bicycle, should not be displaced but managed, and they should be given

Figure 16.2 Solidarity among cyclists – Recreational riders and livelihood bicyclists rally in Lalitpur to demand passage of the Bicycle Act.

Photo credit: Vijay Ratna Shakya.

justice." The Local Government Operation Act (2017) secures the rights of the local government to manage the local market (pp. 11–12). But neither the central nor the local government is visibly working for the benefit of street vendors. They are compelled to continue in their line of work despite the risks to their lives in order to support their families. Apart from life risks from unsafe roads, they are constantly exposed to air pollution. Such problems discourage other people from choosing active transportation like bicycles: but low-wage and marginalized social groups do not have this choice. City planning that emphasizes motorized transportation has little regard for low-income residents (Figure 16.2).

Nepal does have various regulations and policies to manage urban transportation. These include the Transportation Management Act (1993), Road Standards (1998), Transportation Policy (2001), Rural Road Standards (2012), Urban Road Standards (2014), and Provincial Transport Master Plan (2019). But nearly all of these policies favor car culture or motorized vehicles. The electrification of the transport system as per commitment of UNFCCC (Dec 2020) to decarbonize is a hollow promise. Between 2011 and 2016, Nepal's road-transport CO_2 emissions increased by 113% (Maharjan, 2021). Because of the excise duty and tax earned from cars, government is actively promoting car culture.

Local governments' plans like Vision 2035 & Beyond of Kathmandu Valley (2015–2035) have incorporated some provisions of bicycle and pedestrian-friendly infrastructure despite the dominance of motor-centric vision. As part of recent social mobilization, the Cycle Culture Movement coordinated a camp in Dhulikhel on 12–14 March 2023, where participants from diverse fields and policy levels identified five prioritized areas of intervention: (1) the need for Bicycle Act to guarantee cyclist rights and protections, (2) improved bicycle infrastructure, (3) bicycle education from school level, (4) solidarity among bicycle activists and communities, and (5) need of national alliance of bicycle organizations. Together these steps are intended to revive Nepal's bicycle culture. Unfortunately, the majority of participants represented were mountain bike riders – generally from the elite socio-economic class who bicycle mainly for recreation and health. The absence of livelihood-driven cyclists was just noted in the discussion without any substantial reference to their plight.

Conclusion

Although cycling is a crucial element of sustainable transportation (Parkin, 2012), high traffic congestion and pollution in Kathmandu are undermining the health and safety of people who cycle for daily livelihood. At the same time, the contribution of these people to make urban centers like Kathmandu a livable city has not been recognized by the state. They have been excluded and disconnected from policy formulation and are denied fairness and justice. The city planning authorities have not invested in developing infrastructure these people require. Even though they have been contributing to keeping the city clean and livable, there is no proportionate investment in the facilities to protect their health and support their livelihood. The plans and actions of the government are, so far, elite and motor vehicle-centric. The development of Kathmandu has become, as Arundhati Roy writes about globalization more generally, "…like a light which shines brighter and brighter on a few people and the rest are in darkness" (cited in Bunting, 2001). It is a "slow violence" not viewed as violence at all (Nixon, 2011: 2).

Environmental justice demands a different model of urban development. As this study has shown, people who cycle for livelihood are key contributors to the sustainability, vitality, and equity of Kathmandu's urban environment. Accordingly, the unmet needs for safe and healthy transportation for these largely ignored informal sector workers and livelihood cyclers should be incorporated in government's policies and programs for a fair and just city.

Notes

1 Interviewed on April 8, 2023.
2 Interviewed July 21, 2023.
3 FGD & KII in Sanyukta Sarsaphai Jagaran office in Teku.
4 Interviewed on September 17, 2023 in Kalimati.
5 Interviewed on September 21, 2023 in Babarmahal.
6 Interviewed on March 22, 2023.
7 Source: SSP of Nepal Police Sanjeev Sharma (August 16, 2023).
8 Interview in SASAJA office in Teku on July 15, 2023.
9 Interview and FGD in Babarmahal on 10 June 2023.

References

Acharya, A. (2023). Cyclists in Strike after Pathao Reduced Delivery Rate. *Ekagaj*, July 9. https://ekagaj.com/article/society/145337/

Acharya, S. (2022). Pedaling as an Act of Rebellion. *The Record*, February 25. https://www.recordnepal.com/pedaling-as-an-act-of-rebellion

Adetoyinbo, A., et al. (2022). *Technology, Transport, Sustainability: The Undefined Prospects of Lagos Ride*. CEMRI.

Bhattarai, S. (2018). No City for Cyclists. *Nepali Times*, May 12. https://www.nepalitimes.com/banner/no-city-for-cyclists

Bunting, M. (2001). Dam Buster Fiction Review. *The Guardian*, July 27. https://www.theguardian.com/books/2001/jul/28/fiction.arundhatiroy

Dadashova, B., et al. (2023). Equitable Active Transport. *Transport and Environment*, 119. https://doi.org/10.1016/j.trd.2023.103737

Dahal, A. (2023). History of Classic Cars in Nepal. *Nepal Database*, April 27. https://www.nepaldatabase.com/history-of-classic-cars-in-nepal

Gautam, S. (2012). Pedaling for Prosperity. *Nepali Times*, October 5–11. http://archive.nepalitimes.com/news.php?id=19700#.Wu_BkO-FO70

Ghimire, R. (2023). Mobile Street Vendors in Kathmandu – the Local Govt Terrorizes them. *Onlinekhabar*, August 4. https://english.onlinekhabar.com/mobile-street-vendors-kathmandu.html

Haupt, A. (2021). Health Benefits of Biking to Work. *Verywellfit*, July 19. https://www.verywellfit.com/health-benefits-of-biking-to-work-5191939

KhaaliSisi. (2023). Our Contribution. *KhaliSisi.com*, July 31. https://www.khaalisisi.com/contributions_new

Khanal, P. (2021). The Rise and Fall of the Kathmandu Valley's Cycling Culture. *The Record*, June 7. https://www.recordnepal.com/the-rise-and-fall-of-the-kathmandu-valleys-cycling-culture

Lyons, G. (2020). Walking as a Service – Does It Have Legs? *Transportation Research Part A: Policy and Practice*, 137, 271–284.

Maharjan, S. (2021). Nepal's Pathways to Decarbonizing the Transport System. *Blog News*, December13.https://100re-map.net/nepals-pathways-to-decarbonizing-the-transport-system/

Manandhar, T. N. (2013). *Kathmandu...Then and Now.* TN-ND Lumanti Guthi, translated from Nepalbhasha by Razen Manandhar.

Martinez-Allier, J. (2003). *The Environmentalism of Poor. A Study of Ecological Conflicts and Valuation*. Edward Elgar Publication.

NepalLive. (2018). Billions Rupees Go Out of Nepal with Kawadi. *Nepal Live*, August 27. https://nepallive.com/story/30869

Nixon, R. (2011). *Slow Violence and the Environmentalism of the Poor*. Harvard University Press. https://www.google.com.np/books/edition/Slow_Violence _and_the_Environmentalism_o/bTVbUTOsoC8C?hl=ne&gbpv=1

NSO, National Statistics Office. (2023). *National Population and Housing Census 2021 National Report*. NSO.

Ojha, A. (2020). Street Vendors Decry Kathmandu City Office Crackdown on their Businesses. *The Kathmandu Post*, September 18.

Parking, J., ed. (2012). *Cycling and Sustainability*. Emerald Group Publishing Limited. https://books.google.com.np/books?

Pharak Nepal. (2022). Bicycle Lane Should Not Be Constructed Haphazardly. *Pharak Nepal*, June 3. https://faraknepal.com/bicycle-lanes-should-not-be-built-in-hachuwa-mayor-balen-shah

Rai, B. (2011). Cyclists and the City. *Nepali Times*, November 18–24. http://archive.nepalitimes.com/news.php?id=18714#.Wu_AcO-FO70

Regmi, T. B. (2017). A Study of Indian Hawkers in Kathmandu District. Master of Arts in Central Department of Rural Development, Tribhuvan University, MA Thesis. https://opac.tucl.edu.np/cgi-bin/koha/opac-detail.pl?biblionumber=83831&query_desc=au%3A%22Regmi%2C%20Trailokya%20Bikram%22

SASAJA. (2022). *Samyukta Safai Jagaran*. Teku Kathmandu, Mayalbari.
Serkhel, S. (2022). Cycling for Livelihood, Not Lifestyle. Blogs, *Woarchitect*, June 15. https://www.woarchitect.com/cycling-for-livelihood-not-lifestyle
Sharma, S. (2023). Cyclists' Current Road Accident Data Presented in the Workshop organized after Tundhikehl to Dhulikhel Bicycle Mass Rally on May 20.
Shrestha, T. L. and Shrestha, B. (2022). Attitude Towards Bicycle Infrastructure in Nepal with Reference to Hetauda City. *Journal of Humanities & Social Sciences*, 5(3), 275–282. https://www.opastpublishers.com
Sindhu, K. D. (2023). *Roads to the Valley*. Notion Press.
Subedi, A. (2023). Keynote in Cycle Culture in Painting Workshop in Nepal Tourism Board on 30 September. https://arthadabali.com/2023/10/05/49772
Tamang, P. (2022). Cycle Rally (Patan-Darbarmarga) CCM. *YouTube*, June 9. https://www.youtube.com/watch?v=tEfHTn0FmJ8
Wagle, A. (2021). Motorcycle City. *The Kathmandu Post*, April 26. https://kathmandupost.com/columns/2021/04/26/motorcycle-city
Wood, N., and Roelich, K. (2019). Tensions, Capabilities, and Justice in Climate Change Mitigation of Fossil Fuels. *Energy Research and Social Science*, 52, 114–122.

17 Transport Justice on the Streets of Kathmandu

Bhushan Tuladhar

Introduction

Urban planners and managers often fail to realize that people, and not cars, need to be at the center of transport planning. The result is cities that prioritize freeways and flyovers over footpaths, parking lots over public parks, and cars over people. This vehicle-centric growth paradigm, which is common in many cities around the world, including in Nepal, leads to environmental problems, as well as inequities within societies. Low-income and marginalized communities suffer more from inadequate access to transport services as well as disproportionally higher impacts of unsafe streets and air pollution caused by transport systems.

While some cities around the world are starting to invest in people-centric and sustainable modes of transport, such as walking, cycling and public vehicles, cities in Nepal continue to invest in roads and other infrastructure and services that cater to the needs of automobiles. This chapter analyzes existing literature and data on transport systems to assess how this trend is undermining environmental justice, particularly in Kathmandu Valley, Nepal's biggest urban center. It also considers four case studies from Kathmandu – a donor-led road project with little participation from the government and community; a transport project that tried to promote sustainable alternatives; a community initiative to demonstrate a disabled-friendly street; and a local initiative to promote cycling – to analyze what works and what does not in Kathmandu.

Transport Justice Is Environmental Justice

Advocates of transport justice argue that governments have the fundamental duty to ensure that all people from all communities have access to safe, reliable, affordable, and equitable transport systems that connect them to the places, people, and resources they need to enjoy a good quality of life and thrive (Martens, 2017; NACTO, 2016). While this view stresses access to adequate transport services for all, others add that transport justice, like environmental justice, needs to address the unequal distribution of burdens, such as air pollution and noise, and benefits like access to opportunities and facilities, across all communities and places (Karner et al., 2020). They also emphasize that transport justice calls for allowing communities and stakeholders to participate in transport-related planning and decision-making processes (Karner et al., 2020).

DOI: 10.4324/9781003371175-22
This chapter has been made available under a CC-BY-NC-ND 4.0 license.

A transportation system that prioritizes justice is one that prioritizes environment friendly modes of transport such as walking, cycling, and public vehicles and makes these alternatives more accessible, reliable, affordable, and safe for all, particularly for marginalized communities, including women, children, elderly, urban poor and people with disabilities. Such transport systems not only help mitigate social disparities by providing access to improved mobility for all, but they also contribute toward creating a cleaner and healthier environment for all by reducing vehicular emissions and pollution, particularly for vulnerable groups.

Transport Justice and Environmental Justice in the Context of Nepal

Until recently, Nepal has primarily been a rural country. With the exception of Kathmandu Valley, where an urban culture has thrived for centuries, and a few other cities, most Nepali people have lived in scattered villages where mobility needs were often met on foot. In recent years, however, with rapid urbanization and the expansion of a road network, more people have turned to motorized transport such as motorcycles and three-wheelers and to some extent cars.

Nepal has witnessed rapid urbanization along with a growth of transport infrastructure in the past few decades. The total length of the Strategic Road Network (SRN) within the country almost tripled in two decades from 4,740 km in 1998 to 13,448 km in 2018 (DoR, 2018). The SRN consists of highways and feeder roads but does not include local roads and internal urban roads. Therefore, the total increase in road length must be much higher as the percent of population living in urban municipalities increased from 17.1% in 2011 to a whopping 66.17% in 2021 (NSO, 2023). However, much of this increase in length of roads is due to rapid expansion of earthen roads which extended from 179 km in 1998 to 4,191 km in 2018 – an increase of more than 22 fold (DoR, 2018). This may result in more air pollution and road accidents.

Increasing urbanization and road infrastructure has also led to a rapid increase in number of vehicles on the streets, particularly private vehicles. Over the past ten years, the growth rate in motorized vehicles has exceeded 14% per year. Of the 5.26 million vehicles registered in the country up to mid-March 2023, over 81% are motorcycles, and over 6% are cars, while only about 1% are buses. Within the total vehicle fleet registered in the country, the share of public transport (buses, minibuses, microbuses, and three-wheelers) dropped from 11% in 1990 to less than 5% in 2023, indicating a rapid shift toward private vehicles (MoF, 2023).

Even though the number of vehicles has grown significantly over the past few decades, the overall percentage of households in Nepal that own vehicles is still low. The 2021 census indicated that only 3% of Nepali households own a car, jeep, or van while 27% own motorcycles or scooters. In comparison, over 35% of households have bicycles (NSO, 2023). Thus, there is still time to avoid the prominence of car culture in Nepal's cities.

Despite the large number of households with bicycles or no personal vehicles, the government has not introduced programs to ensure the safety of cyclists or improve public transport. Throughout the country, most of the road space is given to vehicles, while cyclists and pedestrians are forced to risk their safety or depend on inconvenient public transport systems. Even in Kathmandu Valley, which is home to roughly 10% of Nepal's nearly 30 million population, sustainable and active mobility options such as walking, cycling, and public transport receive very little investment or attention from the government. Even when resources have been available for improving sustainable mobility options, they have not been effectively utilized (see Box 17.1).

Box 17.1 Kathmandu Sustainable Urban Transport Project – A Lost Opportunity to Promote Transport Justice

The Asian Development Bank (ADB) supported Kathmandu Sustainable Urban Transport Project (KSUTP) was implemented from 2011 to 2018 to address problems such as congestion; inadequate mobility options, particularly for the poor; inadequate transport service; and poor air quality due to transport. The project had four components – public transport improvement; better traffic management; pedestrianization; and cleaning up the air – all of which would have a direct implication on enhancing transport justice in Kathmandu Valley. However, in spite of its noble objectives, the project utilized only 40.6% of its USD 30.42 million budget and achieved only four of its 21 targets. Not surprisingly, the project was labeled "unsuccessful" by the ADB (2020).

The Project Completion Report noted the weak ownership of the project executing agency and inability of the government to implement the plans prepared by the project as the main reasons for its failure. Following the KSUTP, there has been no other major initiative by the government or development partners to promote sustainable mobility project and transport justice.

Evolution of Kathmandu's Transport System

Kathmandu and the other traditional towns in the Kathmandu Valley were designed to facilitate safe and convenient mobility on foot. The houses were built around "*chuka*" or courtyards, which were connected to one another through public passageways under private houses or narrow lanes (*gallis*) in between houses. Walking was convenient and safe. Although low-caste people often lived on the edges of the town, because the towns were compact and neighborhoods were accessible through lanes, mobility posed little problem. In the late 19th and early 20th centuries, however, the Rana rulers of the Valley built vast palaces far outside the traditional settlements and linked them with roads for horse carts. After the 1934 earthquake, wide roads, such as Juddha Sadak, were built.

The first cars were carried by porters into the Valley starting in the late 1930s and early 1940s and used by the Rana rulers. However, because cars were beyond the reach of average citizens and few felt the need for motorized transportation, the number of vehicles remained low. While the rich and powerful could access cars, public transit and non-motorized methods prevailed for the majority of residents.

The first public bus service started in Kathmandu Valley in 1959 by Nepal Transport Service, a private company. In 1961, Sajha Yatayat, a cooperative formed by the government, added more buses and routes. In 1975, electric trolley buses provided by the Chinese government started operating along a 13 km route connecting Kathmandu to Bhaktapur. While these systems provided fairly good service, they were eventually closed by 2010 mainly due to mismanagement and overstaffing. Since then, the government has made very little investment to improve Kathmandu's public transport system, while the private sector has stepped in to partially fill the void. Today, public transport service is operated mainly by the private sector and Sajha Yatayat, which restarted its operations in 2013.

Studies done in 1991 and 2011 indicate that the percentage of trips made on foot and bicycle decreased significantly over two decades while motorcycles and scooters almost tripled during the same period. That the percentage of trips made on public transport has remained more or less constant over the two decades between 1991 and 2011 indicates that, despite little investment in the public transport system, it is still being used by many, particularly those who lack access to private vehicles (DOR/JICA, 2012).

Vehicle-Centric Transport Planning

In recent years, urban transport systems have become more vehicle-centric rather than people-centric as authorities focus on building infrastructure for cars, while ignoring more sustainable and people-friendly alternatives such as walking, cycling, and public vehicles. This trend perpetuates environmental injustice in societies as the majority of people who do not have access to private vehicles are left with few safe options to meet their transport needs. The recent expansion of Ring Road is an example of this trend and its impacts (see Figure 17.1 and Box 17.2).

Figure 17.1 Kathmandu's car-centric Ring Road.

Source: Bhushan Tuludhar.

Box 17.2 Vehicle-centric Expansion of Kathmandu's Ring Road

Kathmandu's Ring Road, a 27 km loop encircling the cities of Kathmandu and Lalitpur, was built in 1977 as a two-lane road. However, with continued urban expansion, the two metropolitan cities have expanded beyond Ring Road. At present, the Ring Road itself is prone to severe traffic congestion.

From 2013 to 2018, a 10 km stretch of the southern section of Ring Road from Koteshor to Kalanki was expanded to 8 lanes – four center lanes and two service lanes on each side – and classified as an urban arterial road. The classification itself was a mistake as per the Indian standard, an urban arterial road is intended for traffic with a design speed of 80 km per hour. Nepal does not have any standards for urban arterial roads. Although the design speed of the expanded Ring Road is only 50kmph, it is still too high for a street passing through dense settlements.

The expanded Ring Road lacks adequate facilities for pedestrians, cyclists, and public transport. The construction agreement signed on 18 December 2012 between the DoR and the Shanghai Construction Group clearly states that the project should include "bicycle lanes and sidewalks on both sides" (Tuladhar, 2021). But this was abandoned for reasons not made public by the government. All eight lanes were given to vehicles with no bicycle or bus lanes. In some areas, there are no footpaths and even where footpaths exist, they are often nothing more than a covered side drain not suitable for walking. Also missing are safe crossings for pedestrians with curb extensions and refuge islands, adequate lighting, medians, street furniture, or traffic calming measures, all of which are necessary for urban streets. As a result, the expanded Ring Road has become one of the most dangerous roads in Kathmandu with frequent crashes claiming many lives. Police report shows that road accident in urban centers like Kathmandu and on the highways is a major killer of youths in Nepal.

While the expanded Ring Road was being designed and constructed, many people and groups raised their voices against it, advocating for a complete street that meets the needs of all road users, particularly pedestrians, cyclists, and public transport users. A group of cyclists even constructed a model lane that went through the existing greenbelt alongside the Ring Road. Sadly, these voices were ignored. Even government agencies that were designing a Bus Rapid Transit in Ring Road were not consulted. Financed by the Chinese Government, the Ring Road expansion was designed with little involvement from Nepali experts or the local community. When confronted by activists, DoR officials said that they did not have detailed designs or plans and once the contract was awarded nothing could be changed.

Public Transport in Kathmandu

Although the private sector has invested in thousands of buses, mini-buses, micro-buses, and three-wheelers to provide public transport services in Kathmandu Valley, the system as a whole is chaotic and inefficient in the absence of government investment and lack of effective monitoring and regulation. A survey of public transport vehicles in Kathmandu showed that overcrowding, excluded passengers, and fleet malfunction were daily occurrences and that public transport vehicles were old and insufficient (DoR/JICA, 2012). The urban poor and others who do not have access to private vehicles are forced to use this ineffective system to meet their mobility needs. The introduction of electric buses could improve the public transport system, but this requires an effective public-private partnership with incentives for private operators to invest in electric buses and government investment in charging infrastructure.

Studies show that overcrowded public vehicles are more inconvenient and unsafe for women. A 2012 survey found that 60% of women (compared to 49% of men) perceived public transport as unsafe at night and 62% of women (compared to 56% of men) felt uncomfortable in public space because of overcrowding (Udas, 2012). In another survey, twice as many women compared to men mentioned being faced with personal insecurity (e.g., fear of pickpockets, personal injury as well as various forms of sexual harassment) while riding public transport and women aged 19–25 years were nine times more likely than men of the same age group to identify personal insecurity as a concern. One in four women and one in ten men who mentioned insecurity as an issue directly experienced 'inappropriate touching' while using public transport (World Bank, 2013).

Non-motorized Transport in Kathmandu

Although Kathmandu was originally designed for walking, it is no longer pedestrian-friendly. While the average trip distance in the Valley is a very walkable 5 km, in the absence of appropriate infrastructure such as sidewalks, pedestrian crossings, traffic calming measures like shade from hot or wet weather, and resting places for pedestrians, walking is inconvenient and unsafe, particularly for children, elderly and people with disabilities.

A study on walkability in Asian Cities showed that Kathmandu is one of the least walkable cities in Asia with a walkability rating of 40 out of 100 (CEN/CANN, 2013). The commercial and public transport terminals in the city, which have the most pedestrian movement, were found to be less walkable compared to residential and educational areas, mainly because of poor pedestrian infrastructure. Walking in Kathmandu is also dangerous: data from the Traffic Police show that pedestrians account for up to 49% of all road fatalities (CEN/CANN, 2013).

Cycling in Kathmandu has also decreased without safe and convenient facilities such as cycle lanes. Bicycles were introduced in Kathmandu Valley in 1903

(Manandhar, 2013), and although initially they were only used by wealthy people, in the 50s and 60s bicycle use also spread to middle- and lower-income people. In the past few decades, bicycle use has decreased (DoR/JICA, 2012) mainly because of the lack of safe cycling infrastructure. However, in recent years, there have been some efforts by Lalitpur Metropolitan City to build bicycle lanes and the number of cyclists has also started to increase with youths taking up cycling as a recreational activity (see Figure 17.2).

Women in particular have also started teaching each other cycling (see Box 17.3).

Figure 17.2 Transport justice: Electric bus and bicycle lanes on the streets of Lalitpur.

Source: Bhushan Tuludhar.

Box 17.3 Women Cyclists and Municipality Leading the Way to Establish a Cycling Culture in Lalitpur

During the second Covid-19 lockdown in 2019, a local women's group in Lalitpur called '*Mahilakaa lagi Mahila*' (Women for Women) started cycling lessons for women. Every morning they gathered in the open grounds of Jawalakhel to teach each other cycling using a few donated bicycles. Many of the initial trainees were female street vendors who lost their income due to the Covid lockdown but over time many others also joined, including homemakers. Over the past three years, the campaign claims to have trained over a 1,000 women to ride bicycles. The campaign has now spread to other parts of Kathmandu where they are similarly training women to cycle.

Many of the women who have been trained say that bicycling has helped them gain self-confidence and also given them a convenient option to move around independently. Cycling has allowed some to expand their businesses such that they now home deliver their products (Awale, 2023). Kathmandu and Lalitpur metropolitans are also helping to expand this initiative.

The Lalitpur Metropolitan City has also joined hands with Nepal Cycle Society, a civil society organization promoting cycling, to design and construct bicycle lanes. In 1999 it built a 5km shared bicycle lane along a major street from Kupundol to Mangalbazar. It then expanded the bicycle lane network to 22 km. Although Lalitpur's bicycle lane is not a new one but rather a part of the existing street that has been marked for bicycles it is a good start nonetheless.

Transport Justice for People with Disabilities

Transport injustice is even more dire for people with disabilities (PWD). The 2021 census indicates that 2.2% of Nepal's population have some form of disability. However, the Nepal Demography and Health Survey (MoHP, 2022) found that among the household population age 5 or older, 23% have some difficulty in at least one of the six core functional domains based on WHO standards. For these people, mobility on Kathmandu's streets is very difficult and at times dangerous, thus denying them access to services and opportunities to experience and enjoy the city.

An accessibility audit by the National Federation of the Disabled Nepal of 150 public infrastructures, including government buildings, public parks and open spaces, roads and streets, corporate sectors, commercial sectors, and other infrastructure within Kathmandu Valley, found that 132 of them (88%) were inaccessible, whereas 18 (12%) were partially accessible, and none were fully accessible (NFDN, 2018).

Tuladhar (2018) demonstrates that the three major challenges faced by PWDs on the streets of Kathmandu are the lack of disabled-friendly infrastructure, people's attitudes and behavior, particularly those of public transport drivers and conductors, and access to information on the transport system. In a survey of PWDs on their transport needs, 77% said that they had to travel outside their houses at least once a day and most of them walk, use a wheelchair, or public transport. Only 6% said they used a car or taxi and 12% said they used their disabled-friendly scooters with four wheels. 93% said that ease of getting on and off the buses was important or very important. 92% said that safety while crossing the street is important or very important.

On many occasions, PWDs have raised their voices demanding accessible facilities in public buildings and spaces by organizing rallies and meeting government officials. At times they have even mobilized themselves to destroy infrastructure that was not disabled-friendly (Regmi, 2019). Even when they have mobilized resources to demonstrate what an accessible street would look like, they have faced many challenges in implementing their design (see Box 17.4).

Box 17.4 Community Efforts to Demonstrate an Accessible Street in Jorpati

In 2011, Khagendra Navajeevan Kendra, a center for people with disabilities, together with Spinal Injury Rehabilitation Centre, decided to mobilize local resources to convert a 100 m stretch of a street in Jorpati area in Kathmandu into an accessible street in order to demonstrate how existing streets can be made disabled-friendly. They organized a series of discussions, co-designed the street with the help of an architect, raised funds locally, and convinced the Ministry of Physical Infrastructure and Transport to expand the stretch to 1 kilometer.

The project was expected to be completed within a year and benefit 5,000 local people with disabilities. However, the community faced a lot of resistance from different authorities as well as some locals while expanding the sidewalks, shifting electricity poles, removing illegal structures, and getting the contractor to follow quality standards and deadlines. The project was finally completed in 2023, although many of the intersections with side roads are still to be fully completed.

Conclusions and the Way Ahead

Because of a vehicle-centric planning model adopted by government authorities in Nepal as well as international agencies, transport justice has become a key issue of concern in Kathmandu. Many people, particularly the urban poor, women, children, elderly, and people with disabilities, face difficulties in accessing adequate transport services to meet their mobility needs. They also face the brunt of the adverse impacts of transport such as air pollution and unsafe roads. A shift to more sustainable modes of transport such as walking, cycling, and electric public transport can lead the way to improved environmental conditions as well as transport justice on the streets of Kathmandu.

Several case studies of recent transport-related initiatives indicate that a shift to more environment-friendly transport system that ensures justice for all in Kathmandu Valley and beyond is possible. But it will require the government, particularly the Ministry of Physical Infrastructure and Transport, and municipalities, to take the lead in designing and implementing people-centric transport systems. A crucial component of this initiative is the participation of all stakeholders and active engagement of local communities, including the urban poor, women, children and the elderly.

References

ADB (2020). Completion Report Nepal: Kathmandu Sustainable Urban Transport Project, Asian Development Bank, Manila. 44058-013: Kathmandu Sustainable Urban Transport Project (adb.org).

Awale, S. (2023). Life Cycle. Nepali Times. 8 October 2023. Kathmandu. https://nepalitimes.com/multimedia/life-cycle

CEN/CANN (2013). Walkability in Kathmandu Valley. MaYA Factsheet #2. Clean Energy Nepal, Clean Air Network Nepal and UN-Habiat. Kathmandu. https://www.cen.org.np/public/uploads/doc/walkability-in-kv-maya-factsheet-2-60b9f655159d9.pdf

DoR/JICA (2012). Data Collection Survey on Traffic Improvement in Kathmandu Valley: Final Report. Department of Roads Ministry of Physical Planning, Works and Transport Management. Japan International Cooperation Agency 2012. http://open_jicareport.jica.go.jp/pdf/12082459_01.pdf

DoR (2018). Statistics of Strategic Road Network, Department of Roads, Ministry of Physical Infrastructure and Transport, Government of Nepal, Kathmandu. https://dor.gov.np/home/publication/statistics-of-strategic-road-network-2-17-18/force/part-2-ssrn-2-17-18

Karner, A., London, J., Rowangould, D., & Manaugh, K. (2020). From Transport Equity to Transport Justice: Within, Through, and Beyond the State. *Journal of Planning Literature* 35 (4). https://journals.sagepub.com/doi/10.1177/0885412220927691

Manandhar, T. N. (2013). Kathmandu Then and Now. Tirtha Narayan-Narayan Devi Lumanti Guthi. Kathmandu. https://shopratnaonline.com/kathmandu-then-and-now-tirtha-narayan-manandhar/

Martens, K. (2017). *Transport Justice: Designing Fair Transport System* (1st Edition). Routledge.

MoHP (2022). Nepal Demography and Health Survey 2022: Key Indicators Report, Ministry of Health and Population. New Era. ICM. Kathmandu. https://mohp.gov.np/uploads/Resources/Nepal%20Demographic%20and%20Health%20Survey%202022%20Key%20Indicators%20Report.pdf

MoF (2023). Economic Survey 2079/80. Ministry of Finance. Government of Nepal. Kathmandu. https://www.mof.gov.np/site/publication-detail/3248

NACTO (2016). Global Street Design Guide. Global Designing Cities Initiative. National Association of City Transportation Officials. Island Press. https://nacto.org/publication/global-street-design-guide/

NFDN (2018). Report on Accessibility Audit in Kathmandu Nepal. National Federation of the Disabled Nepal in partnership with CBM and Kathmandu Metropolitan City. Kathmandu. Final_Access-audit-report_NFDN2018.pdf

NSO (2023). National Population and Housing Census, 2021 National Report. National Statistics Office, Government of Nepal. Kathmandu. https://censusnepal.cbs.gov.np/results/downloads/national

Regmi A. (2019). People in Wheelchairs Pull Down Recently Built Sidewalks, Demand Disabled-Friendly Infrastructure. The Kathmandu Post. March 28, 2019. Kathmandu. https://kathmandupost.com/valley/2019/03/28/under-construction-sidewalk-demolished-to-protest-lack-of-disabled-friendly-infrastructure

Tuladhar, B. (2018). Access to Transport Services for People with Disabilities in Kathmandu. The Prospects and Situation of Accessibility in Nepal A Collection of Articles. National Federation of the Disabled – Nepal and CBM International. Kathmandu. https://nfdn.org.np/wp-content/uploads/2019/08/Final_article-collection-on-accessibility_NFDN2018.pdf

Tuladhar, B. (2021). Cycle to Fight Climate Change. Nepali Times. 31 October 2021. Kathmandu. https://nepalitimes.com/opinion/comment/cycle-to-fight-climate-change

Udas, S. (2012). Public Transport Quality Survey. Kathmandu: Clean Energy Nepal and Clean Air Network Nepal. https://cen.org.np/clean-air-and-urban-mobility/public-transport-quality-survey.html

World Bank (2013). Riding a Bus in Kathmandu: Gender and Transport in Nepal. Kathmandu. https://www.worldbank.org/en/news/feature/2014/03/17/riding-a-bus-in-kathmandu-gender-and-transport-in-nepal

18 Building Political Capabilities through Participation for Environmental Justice in Informal Housing in Kathmandu

Sangeeta Singh and Bijay Singh

Introduction

Squatter settlements, in which the inhabitants do not have official land ownership rights, are a common phenomenon in Nepal and other parts of the Global South. Residents of informal settlements – sometimes called *sukumbasi* in Nepal – face persistent challenges such as being denied access to basic infrastructure and services. The location of these settlements alongside river floodplains and steep slopes exacerbates their vulnerabilities to various environmental hazards, including flooding, landslides, air pollution and water contamination that pose severe health risks. The disproportionate burden of environmental hazards and pollution on informal settlements, particularly those inhabited by marginalized and vulnerable populations, makes a case for environmental injustice. Squatter communities often lack the political capital to access essential resources. Moreover, their informal land tenure leaves them vulnerable to eviction threats, as they lack legal ownership of the land. Consequently, residents of these settlements live in precarious conditions, with their homes and livelihoods constantly at risk and their very existence stigmatized. The eviction of squatter settlements is a contentious issue, often giving rise to various social, legal, and human rights concerns.

Government policies frequently disregard the specific challenges encountered by informal settlements, revealing a gap in acknowledging and addressing environmental injustices. The prioritization of top-down and standardized housing solutions, driven by technocratic rationality, often fails to account for the diverse needs and contexts of informal settlements. An illustration of this is evident in the Government's urban poor housing program of 2012. In this initiative, the Department of Urban Development and Building Construction, operating under the Ministry of Urban Development, procured 4,070 square meters of land from the Ichangu Narayan Land Pooling Committee. Subsequently, in 2014, they constructed 227 housing units at a cost of Rs230 million for squatters and urban poor in informal settlements along the Bagmati River in Kathmandu Valley. Despite these efforts, the housing complex has remained vacant for nearly five years as the squatters are unwilling to relocate (Chand, 2019). Currently, the housing is being repurposed for a different use.

DOI: 10.4324/9781003371175-23
This chapter has been made available under a CC-BY-NC-ND 4.0 license.

In Nepal and other parts of the Global South, informal settlers persistently advocate for better infrastructure by leveraging their affiliations with various civil society organizations, social movements and drawing on historical experiences of collaboration. They make their demands through strategic approaches aimed at accessing infrastructure. These communities demonstrate remarkable resilience and solidarity, forging strong social networks to navigate challenges. They see grassroots initiatives as a solution to their problems, actively participating in building the necessary infrastructure (Tanaka 2009).

This chapter presents a retrospective study of the Kirtipur Housing Project in the Kathmandu Valley, focusing on the evolving process of participation and its contribution to political capabilities. It explores how participation in consensus-building and contention across communicative forums enhances knowledge, reshapes political networks, and influences discourse. This research uses key informant surveys, observation, ethnography, and literature reviews. The first author participated in both the preparatory and design phases, as well as in the construction phase of the project.

The premise of the research is grounded in the belief that involving the communities in decision-making processes through participatory methods leads to enhanced political capabilities which in turn leads to environmental justice.

Theoretical Framework: Political Capabilities, Participation and Environmental Justice in Informal Housing

Environmental justice encompasses movements and discussions aimed at addressing environmental inequalities, reducing risks, and enhancing protections, particularly for minority and economically disadvantaged communities (Khosravaninezhad & Akbari, 2014). A capabilities approach, popularized by Amartya Sen, prioritizes the enhancement of individuals' capabilities to lead meaningful lives, emphasizing equality through equalizing opportunities and freedoms, as opposed to a narrow focus on material resources (Sen, 1979, 1984; Edwards 2016). Sen contends that people have not only a right to basic resources but also to the freedoms and opportunities that enable them to convert these resources into valuable functionings and capabilities (Sen, 1984). This framework places individuals' capabilities at its core, taking into account the influence of their social, economic, and cultural contexts (Sen & Nussbaum, 1993). Nussbaum (2011) identifies specific capabilities as inalienable human rights, such as life, physical well-being, emotional health, practical reasoning, and social connections, and emphasizes the roles of governments and communities in creating environments that enable individuals to fully realize their capabilities. Schokkaert's political capability method (2008a, 2008b) delves into the normative aspects of political engagement, stressing on sustainable involvement and providing information to marginalized individuals. Turner's classic work (1976) advocates a housing approach centered on community participation and self-help to address housing challenges and supports a bottom-up, participatory approach where residents actively shape their living environments. This approach harnesses local knowledge, collective decision-making, and social networks to create sustainable housing solutions, asserting that communities can more effectively meet their unique needs when actively involved in shaping their housing environments.

The Case of Relocation of Informal Settlement: Kirtipur Housing

Informal Housing in Kathmandu Valley

The growth of the urban population in Kathmandu which accelerated in the 1990s due to people fleeing the Maoist insurgency in the hinterlands put a drain on limited employment opportunities available in the agricultural sector. People also migrated to Kathmandu in search of better economic and social opportunities. This in turn also contributed to the increase in the number of informal settlements throughout Kathmandu Valley. We do not have the latest figures but as of 2010, there were approximately 75 informal settlements in the Valley, which is a significant increase from the 17 recorded in 1985 (Toffin, 2010). According to a survey conducted in 2013 by two organizations representing informal settlers, there were 24,021 squatters residing in 46 informal settlements in Kathmandu (Dangol & Day, 2017). A 2017 report from the Nepal Landless Democratic Union Party counted the number of squatter families in Kathmandu at 29,000 across 73 settlements (Ojha, 2017).

These settlements are primarily situated along riverbanks, while a smaller number can be found in non-riparian environments, particularly in the suburbs of Kathmandu, often on the periphery of core settlements. The concentration of squatters along the banks of rivers is attributed, in part, to the fact that the ownership of these riverbanks lies with the central government rather than the municipality (Toffin, 2010). Marginalized communities in informal settlements face significant challenges in gaining access to basic services and amenities, and often form informal networks and self-organized groups to tackle infrastructure issues collectively.

Non-Governmental Organizations (NGOs) and community-based organizations often play a crucial role in supporting these communities, providing assistance in securing funds, technical expertise, and through advocacy to improve infrastructure. While formal government support is often limited, there are cases where local governments recognized the importance of tackling infrastructure challenges in informal settlements. In these instances, they have taken steps to provide resources, such as electricity.

The Kirtipur Housing Project

The Kirtipur Housing Project in Kathmandu sought to address the eviction of informal settlers along the Vishnumati Corridor due to road construction by the Government in 1980. By 2005, out of the 142 households that were subjected to eviction, 44 were resettled in formal housing. This resettlement was based on the identification and selection of genuine squatters, excluding the majority.

Despite its shortcomings, the project used a participatory approach, involving affected communities in decision-making, rather than top-down planning.

This ensured that solutions were tailored to the real needs and aspirations of the residents, empowering them in the design and planning of their new homes. Grassroots organizations played a pivotal role by advocating for the rights and needs of marginalized communities. They leveraged their political capabilities to influence policy decisions and secure resources, garnering support from government authorities and stakeholders.

The project illustrates how a combination of participatory approaches, political advocacy, and environmental considerations can benefit marginalized communities. Relocating these communities to safer areas not only improved living conditions but also reduced vulnerability to natural disasters, contributing to environmental justice. Examining the relocation process within the project reveals valuable insights for achieving sustainable urban development while supporting marginalized communities.

The Eviction Process

The eviction strategy employed in the Bagmati River area was a component of the "Integrated Development of the Bagmati Civilization (IDBC)" project. This initiative was motivated by the significant pollution resulting from informal settlements situated in close proximity to the riverbanks (Shrestha et al., 2014). The Vishnumati Link Road Project, initially conceived in 1969, aimed to create a north-south connection (Sorakhutte-Kalimati) along the Vishnumati River Corridor. It was expected to impact five communities and entail the demolition of 142 houses that were located in Dhukhal, Chagal, Kushibahil, Tankeshwar, and Dhaukhel on the right side of the river. The oldest settlement, Tankeshwar, dating back to 1952 and the latest, Chagal, dating back to 2000 (Toffin, 2010). The road project faced numerous planning stages and funding challenges, including the Asian Development Bank's (ADB) withdrawal in 1999–2000 due to the government's refusal to compensate and resettle squatters affected by the road construction. Construction only began in 1999, with eviction notices issued to affected residents and bulldozers deployed in 2002 to demolish structures. Families took various actions, from voluntarily dismantling their homes for compensation and permanent housing to relocating to Kathmandu or nearby areas while some had no alternative and resettled in the same area.

At the time, the prevailing discourse viewed these settlements as informal and their residents as illegitimate settlers. Studies between 1992 and 2002 emphasized the need for providing affordable alternative housing before dismantling existing homes, in alignment with Nepal's Constitution and Shelter Policy. A significant development occurred when City Development Strategy (CDS) of the Kathmandu Metropolitan City (KMC) in 2001 involved squatter community members in crafting policies and strategies to improve living conditions and provide tenure security, adopting a participatory approach. The Supreme Court upheld the government's eviction decision but also ordered appropriate alternatives for the squatter settlement dwellers, highlighting the importance of balancing development goals with human rights protection.

The Development of Alliances and Networks

In 1999, Kalimati residents sought government compensation and Lumanti, an organization focused on housing and poverty, joined their efforts, engaging in discussions, surveys, documentation, and alternatives exploration. Despite their work, a second notice in January 2002 pushed for settlement demolitions. Meetings involving residents, Lumanti, legal representatives and officials, the Vishnumati Link Road Project team, the Department of Urban Development and Building Construction (DUDBC), and squatter federations (Nepal Basobas Basti Samrakshan Samaj and Nepal Mahila Ekta Samaj) temporarily delayed construction, but a subsequent eviction notice in March 2002 gave residents just three days to vacate their homes.

During this period, Lumanti actively lobbied with various government officials, including the secretary and minister of the Ministry of Physical Planning and Works, the DUDBC director, and the Mayor of Kathmandu Metropolitan City (KMC), urging them to explore alternatives before proceeding with the demolitions. Lumanti sought support from various organizations and garnered media attention to delay demolitions until after school exams. They documented the communities through photos and videos, engaging affected families and alliances in their efforts to delay demolitions until viable alternatives were provided. Their advocacy efforts bore fruit when the Mayor of KMC agreed to postpone demolitions and identify genuine squatters for compensation. Communities submitted applications for housing or financial support with Lumanti's assistance, laying the foundation for early advocacy. In a March 23, 2002 meeting, squatters secured a verbal promise from the Mayor of KMC, including temporary rental housing and eventually government-built housing for genuine squatters. The agreement also stipulated financial support, with affected families receiving Rs 2,000 per month for three months to assist with rent, and genuine squatters having the option to purchase houses through monthly instalments.

The ADB's withdrawal from the project due to the government's reluctance to relocate the squatters before development, along with academic research highlighting negative impacts, reshaped the discourse and facilitated dialogues with the government and development agencies. There was a notable shift in discourse that placed greater emphasis on humanitarian concerns and destigmatization rather than solely on property rights. This shift involved highlighting the potential adverse effects on school exams and the future prospects of children. While approaching agencies, NGOs, and INGOs did not yield immediate alternatives, it prolonged demolition dates and enhanced the alliances' reputations. The success of the communities and their alliances in delaying demolition and advocating for fair compensation and alternatives stemmed from a multifaceted approach which included early engagement and advocacy, data collection and documentation, engagement with decision-makers, collaboration with other organizations, media involvement, public awareness, database, and knowledge co-creation, and persistent efforts to shape the discourse and build a strong network.

Identification of Informal Settlers

The communities and their allies faced challenges even after delaying demolition and persuading the government to consider alternative solutions. A key challenge was defining "genuine squatters," which shifted from the conventional definition of unauthorized occupants to those who had no land elsewhere and couldn't afford housing, even rentals. Identifying these genuine squatters required strong community support and trust-building with Lumanti. Verification of the squatter list was complicated by the need for it to be approved by the Ward Committees, and it also needed assessments from the time when the Asian Development Bank (ADB) was planning to fund the project, which meant demolishing houses before compensation. Many residents had lost their official documents and couldn't afford demolition without the promised Rs 2,000 compensation.

The project team and Lumanti worked together to identify genuine squatters through detailed surveys, which collected data on households, income sources, occupations, and house numbers, enabling tailored solutions to meet the community's specific needs.

The Urban Community Support Fund (UCSF)

An Urban Community Support Fund (UCSF) was established to finance the relocation of urban communities, sourcing funds from government budgets, donor organizations, and community contributions. The UCSF operated on cooperative principles, with diverse boards managing funds, ensuring inclusive decision-making. The UCSF aimed to promote housing ownership, financial access for income-generating activities, and capacity-building for urban poor communities. It emphasized supporting groups rather than individuals, adopting a bottom-up, participatory approach, and empowering communities. The establishment of the fund marked a pioneering collaboration between the urban poor and the local government, setting a significant precedent for both Kathmandu and Nepal.

While contributors and supporters, including the Mayor of KMC and Asian Coalition for Housing Alliance (ACHR) Thailand's representative to the ADB, played vital roles, the ownership and management of the fund remained with the community. Community ownership and management of the fund were paramount in maintaining grassroots control. Lumanti acted as the fund's administrator, enhancing transparency and collaboration. The UCSF served as a platform for building community capacity, enabling negotiations with the government and facilitating alliances between the state and federations of the landless and urban poor.

Choice of Housing Sites

A suitable relocation site was chosen through a comprehensive assessment that considered safety, access to services, and environmental sustainability. Community input played a crucial role in the selection process, ensuring alignment with residents' preferences and cultural ties. In June 2002, Lumanti, in collaboration with community members and federations' representatives, worked with the mayor of KMC to explore potential resettlement sites. After considering several alternatives, the site was ultimately selected at Paliphal in Kirtipur, situated 8 km from the center of Kathmandu.

The decision-making processes regarding potential resettlement sites and housing options highlight the importance of community participation. Lumanti and community members actively engaged in site inspections, ensuring that locations chosen were accessible and affordable, aligning with their needs. Active community engagement increased the likelihood of developing healthy, environmentally-friendly and community-responsive housing solutions.

Housing Settlement Planning and Design

The planning and design phase of the project involved close collaboration between architects, urban planners, and relocated squatters. Together, they created a conceptual plan for the new housing settlement, taking into account the layout, housing units, infrastructure, and public spaces. Community input played a significant role in shaping the design, ensuring it met their needs and aspirations.

Funding for the land purchase came from UCSF, along with government subsidies. The selection of the location was based on community preferences and budget considerations, despite some compromises. For example, while the site was somewhat isolated in terms of social networks, efforts were made to mitigate this by involving children in nearby school activities. Provision of motorcycles facilitated residents' commute to their workplaces. A total of 44 two-story houses were built, offering two distinct designs. The construction was highly cost-effective, standing at almost half the price of other buildings with similar characteristics during the time of construction.

The families who were relocated to these houses recognized that they wouldn't have been able to own such housing through personal investments. The project's focus on cost-effectiveness, efficient use of materials and technology, and maximizing available resources resulted in an efficient and sustainable housing solution. The premises also incorporated features to promote community cohesion and ecological sustainability, including spacious open areas for gatherings, rainwater harvesting, and a constructed wetland for decentralized wastewater treatment.

Financing Mechanism

The project accommodated the financial limitations of the residents and provided them with an interest-free loan repayment system. This approach made homeownership a realistic goal for these families. This was a critical step in ensuring that the community could access secure housing. This loan allowed them to acquire housing without the burden of high interest rates, making the process more affordable and sustainable for the community. According to interviews, the families have diligently repaid the loan over 15 years. This commitment to loan repayment demonstrates the community's responsibility and determination to eventually claim ownership of the houses. It reflects the success of the cooperative-based scheme in achieving its objectives.

The Kirtipur Housing Management Committee

The establishment and evolution of the Housing Management Committee (HMC) marked a significant development in representing the urban poor, combining technical expertise with community aspirations and needs.

The Kirtipur municipality and NGOs primarily focused on promoting voluntary membership and participation within the HMC. This approach empowered community members to make decisions about land selection, housing schemes, and the path to homeownership, emphasizing a participatory and community-driven approach. Trust and active participation among community members were generated through voluntary activities and were evident in various project stages, contributing to its feasibility and success.

One challenge of this organizational model was the absence of a formalized policy document guiding its operations, sometimes not fully incorporating grassroots initiatives and community values. This highlighted the need for a more inclusive and community-centered policy framework.

Land Tenure and Ownership

The project formalized land tenure and ownership arrangements to ensure residents had secure rights to their new homes. The initial design intended to grant individual ownership of houses to residents while keeping the land under community ownership. However, challenges arose related to the payment of land taxes, which were typically held collectively by the community. Resolving these tax payment issues became a time-consuming process, further complicating the transfer of land tenure to individual homeowners. The issue of land taxes and the complexities surrounding community-held taxes contributed to setbacks in transferring land ownership to individual residents. This situation underscores the importance of addressing legal and financial matters promptly and efficiently when implementing housing projects with shared land ownership models.

Housing Related Capabilities

The Kirtipur Housing Project has proven successful in enhancing various housing-related capabilities across multiple dimensions, along the lines of what Sen (1980, 1984) and Nussbaum (2011) recommend, including social, environmental, economic, physical, and institutional aspects.

The project's design emphasized fostering social interactions and community cohesion. While initial challenges related to social acceptance due to diverse resident backgrounds existed, the project provided an opportunity for different communities to coexist and gradually build understanding and acceptance. Resistance to resource-sharing with neighboring communities highlights the importance of addressing equitable access arrangements.

In terms of physical capabilities, the project's use of a spacious courtyard as a safe gathering area during emergencies contributes to disaster resilience and community safety. Additionally, the incorporation of ecological measures like rainwater harvesting and decentralized wastewater treatment systems demonstrates a commitment to environmental sustainability, even though challenges related to the operational status of the wastewater treatment system exist.

Economically, despite initial challenges regarding residents' distant workplaces, improved bus transit services later addressed this issue, enhancing economic capabilities by reducing commuting hardships and improving access to employment opportunities. Institutionally, effective planning, coordination, and governance mechanisms were evident in managing the project and addressing residents' needs and concerns.

The successful execution of the housing project has significantly enhanced housing-related capacities across various aspects. Its holistic strategy, centered on fostering community unity, involving participatory planning, prioritizing ecological sustainability, ensuring economic accessibility, bolstering disaster resilience, and reinforcing institutions, has notably elevated the residents' quality of life and well-being, despite encountering obstacles along the journey. Lumanti's CEO emphasized how this project's success has catalyzed further endeavors targeting urban poverty through housing.

> Moving beyond mere advocacy, we ventured into on-site housing development and infrastructure enhancement initiatives. Partnering closely with local municipalities, our aim was to establish housing projects for marginalized communities. The Kirtipur Housing Project acted as our pioneering effort. Our commitment endures in implementing on-site housing construction approaches in other settlements.
>
> (Personal communication 2022)

Conclusion

The challenges faced by informal settlements in Kathmandu are representative of broader issues related to social inequality, environmental injustice, and the complexities of urban development in developing regions, especially in the Global South. The Kirtipur Housing Project's case study serves as a model, showcasing that a combination of participatory approaches, political empowerment, and environmental fairness can bring about significant positive changes.

This case study emphasizes that community participation should not be a superficial gesture but a fundamental component of successful urban development initiatives. Grassroots organizations played a pivotal role in advocating for the rights and needs of marginalized communities, highlighting the importance of political capabilities in influencing policy decisions and securing resources. Additionally, the project's commitment to environmental justice, relocating communities from vulnerable floodplain areas to safer locations, demonstrates a dedication to improving living conditions and reducing the disproportionate burden of environmental risks on marginalized populations.

The Kirtipur Housing Project has succeeded in enhancing housing-related capabilities across various dimensions. Its holistic approach to community cohesion, participatory planning, ecological sustainability, economic accessibility, disaster resilience, and institutional strengthening has significantly improved the quality of life and well-being of its residents. While there were challenges to overcome, the project's positive impact on the community remains a good example of effective urban development. Today, the residents of the settlement have made significant progress, forming a united and cohesive community. Their social interactions with the neighborhood have improved, fostering acceptance. They take pride in being legal homeowners in the community, a hard-won achievement.

Formalization of land tenure, with careful consideration of each settlement's unique context and needs, coupled with community participation and a focus on social justice, can contribute to more inclusive, sustainable, and resilient urban development. For policymakers, urban planners, and development practitioners, this case study offers valuable lessons about inclusivity, community engagement, and thoughtful planning in addressing the intricate challenges of informal settlements.

References

Chand, B. J. K. (2019). Land Management Problems in Nepal: Repercussions of Land Use Policy and Implementation Gaps in Regional and Urban Planning. *ISPRS Annals of the Photogrammetry, Remote Sensing and Spatial Information Sciences*, Volume IV-5/W2, 2019 Capacity Building and Education Outreach in Advance Geospatial Technologies and Land Management, 10–11 December 2019, Dhulikhel, Nepal.

Dangol, N., & Day, J. (2017). Flood Adaptation by Informal Settlers in Kathmandu and their Fear of Eviction. *International Journal of Safety and Security Engineering*, 7, 147–156. https://doi.org/10.2495/SAFE-V7-N2-147-156.

Edwards, G. A. S., Reid, L., & Hunter, C. (2016). Environmental Justice, Capabilities, and the Theorization of Well-being. *Progress in Human Geography*, 40(6), 754–769.

Khosravaninezhad, S., & Akbari, R. (2014). Application of Environmental Justice Concept in Urban Planning, the Peri Urban Environment of Tehran as the Case Study. *International Journal of Architectural Engineering and Urban Planning*, 1, 56–64.

Manandhar, L. (2022). Interview by First Author with Lajana Manandhar, Chief Executive Officer, Lumanti, June 2022.

Nussbaum, M. C. (2011). *Creating Capabilities: The Human Development Approach*. Cambridge: Belknap Press of Harvard University Press.

Ojha, A. (2017). Squatter Settlement Demolished. Kathmandu Post, July 16, 2017. https://kathmandupost.com/valley/2017/07/16/squatter-settlement-demolished.

Sen, A. (1979). Equality of What?, in: S. McMurrin (Ed.), *Tanner Lectures on Human Values*. Cambridge: Cambridge University Press, pp. 196–220.

Sen, A. (1984). Rights and Capabilities, in: A. Sen (Ed.), *Resources, Values and Development*. Oxford: Blackwell, pp. 307–324.

Sen, A. & Nussbaum, M. (1993). *The Quality of Life*. New York: Oxford University Press.

Schokkaert, E. (2008a). The Capabilities Approach. https://ssrn.com/abstract=1084821 or http://doi.org/10.2139/ssrn.1084821.

Schokkaert, E. (2008b). Capabilities and Satisfaction with Life. *Journal of Human Development and Capabilities*, 9(3), 377–392.

Shrestha, R., Tuladhar, A., Zevenbergen, J., & Banskota, M. (2014). “Decades of Struggle for Space”: About the Legitimacy of Informal Settlements in Urban Areas. Paper presented at the XXVInternational Federation of Surveyors Congress. Kuala Lumpur, Malaysia.

Tanaka, M. (2009). From Confrontation to Collaboration: A Decade in the Work of the Squatters’ Movement in Nepal. *Environment and Urbanization*, 21(1), 143–159.

Toffin, G. (2010). Urban Fringes: Squatter and Slum Settlements in the Kathmandu Valley (Nepal). *Contributions to Nepalese Studies*, 37(2), 151–168.

Turner, J. F. C. (1976). *Housing by People: Towards Autonomy in Building Environments*. London: Marion Boyars.

Part 6

Climate Justice

19 Climate Change in Nepal through an Indigenous Environmental Justice Lens

Pasang Yangjee Sherpa

Introduction

Indigenous Environmental Justice (IEJ) is a distinct formulation of environmental justice required to address the challenges of the ecological crisis as well as the various forms of violence and injustices experienced specifically by Indigenous peoples (McGregor 2021). It extends conventional environmental justice scholarship in diagnosing injustice to humanity and seeking remediation outside of the confines of the current political, economic, and legal context (McGregor 2021). It offers the necessary space and distance from which to examine the legitimacy, applicability, and effectiveness of dominant global and nation-state political, legal, and scientific mechanisms. It presents a decolonial way forward. Kyle Whyte's (2021) "epistemologies of coordination," for example, emphasize the moral bonds – of kinship relationships – for generating the responsible capacity to respond to constant change. The coordination framework assesses the impacts of actions by their contributions to the quality of kinship relationships and draws our attention to how problems of *presentism* and *imminence* in "epistemologies of crisis" can betray ethics and justice. Imminence refers to the "sense that something horribly harmful or inequitable is impending or pressing on the present conditions people understand themselves to be living in (Whyte 2021: 54)" Presentism in this context refers to how someone becomes so concerned with the *present* crisis as *new* that they neither question their own perspective, including the presumed neutrality of time, nor where their perspective may derive its social origins (Whyte 2021: 54). The obsession with presentism obscures how everyone else may experience today's world, and the sense of imminence overshadows the realism needed to remember how colonial and other forms of power engendered the current state of affairs and how these forms of power are poised to retrench (Whyte 2021: 61).

This chapter comes on the heels of "Water, ice, society, and ecosystems in the Hindu Kush Himalaya: An outlook," a report produced by the International Center for Integrated Mountain Development (ICIMOD 2023). Along with the Hindu Kush Himalaya Assessment (Wester 2019), this report serves as the authoritative scientific knowledge on climate change impact in Nepal and the HKH region for planning and policymaking. They also feed into the global climate assessment reports such as the Intergovernmental Panel on Climate Change (IPCC) Assessment

DOI: 10.4324/9781003371175-25
This chapter has been made available under a CC-BY-NC-ND 4.0 license.

Reports (ARs) with direct implications on regional and planetary imagination of the present and future (IPCC 2023). The report states that glaciers disappeared 65% faster in the 2010s than in the previous decade and with current emissions pathways, 80% of glaciers' current volume will disappear by 2100. This would mean that the availability of water would peak by mid-century and decline thereafter. It is also reported that vulnerable mountain communities are already experiencing major adverse impacts, and the impacts on fragile mountain habitats are particularly acute.

This chapter takes IEJ perspective to reexamine climate change adaptation in Nepal. In doing so, it recognizes: (1) Indigenous conceptions of justice do not separate environmental concerns from spiritual ties to the land, mountain, air, and water; and (2) social, political, and economic survival of Indigenous communities are vital features of IEJ. The chapter draws upon ethnographic insights on climate change activities and lived realities from the high mountains of Nepal based on fieldwork conducted between 2009 and 2023. The first part of the chapter examines Nepal's core climate policies and the second part examines climate change adaptation at the local level.

Development Aspirations of Nepal's Climate Change Adaptation

An examination of climate change adaptation in Nepal requires taking a closer look at the national climate change policies to understand how national authorities perceive and respond to climate change. Several documents make up Nepal's core climate policy: The *Climate Change Policy*, National Adaptation Program of Action (NAPA), the National Framework for Local Adaptation Plans of Action (LAPA), and the National Planning Commission's (NPC) 'Climate Resilient Planning'. The Approach Paper to the NPC's Thirteenth Plan (FY 2013/14–2015/16) that sets development agenda and approach mentions that the objective of pursuing 'environment and climate change' by the NPC is to 'adapt to the adverse impacts of climate change as called for under the principles of green development' (GoN 2013: 115).

Nepal's National Planning Commission (NPC) prepared its 'Climate Resilient Planning' document to help ministries, departments, and development organizations in analyzing sector-specific climate issues with a greater understanding of climate variables at the local level and in adopting measures to reduce the emerging and anticipated climate threats which development plans and programs face (NPC 2011: 4). The Climate Resilient Planning document envisions achieving a society and economy that is resilient to a changing climate as below:

> All people, including the poor and vulnerable, have the capacity to respond in an adaptive (as opposed to reactive) way to current and future climate risks. They will have many choices, feel secure, and will be willing and able to invest in improving their livelihoods. Formal and informal institutions will reinforce the abilities of individuals to predict, prepare for, and recover from climate shocks. They will learn to monitor and respond to

> changing conditions in a timely, flexible and efficient manner. Practitioners and policy-makers will be equipped with the knowledge, tools, enabling policies and sustained funding needed to implement decision in a manner that increase resilience.
>
> (NPC 2011: 24)

NAPA as a strategic tool assesses climatic vulnerability, and systematically responds to climate change adaptation issues by developing appropriate adaptation measures (MoE 2010). The NAPA project identifies six thematic working groups: (1) agriculture and food security; (2) climate-induced disaster; (3) urban settlement and infrastructure; (4) public health; (5) forests and biodiversity; and (6) water resources and energy. It also gives two cross-cutting themes: Livelihoods and governance, and gender and social inclusion. The National Framework for LAPA document (GoN 2011b) was designed to encourage consultation with local communities and to produce projects relevant at that level. It emphasizes a 'bottom-up, inclusive, responsive and flexible' approach to climate adaptation (Helvetas and RRI 2011: 14). LAPA supports the 'operationalization of the policy objectives outlined in the NAPA, the *Climate Change Policy* and Climate Resilient Planning by facilitating the integration of climate change resilience into local-to-national development planning processes and outcomes' (GoN 2011b: 2).

The *Climate Change Policy* (GoN 2011a: 5) envisions 'a country spared from the adverse impacts of climate change, by considering climate justice, through the pursuit of environmental conservation, human development, and sustainable development – all contributing toward a prosperous society'. The main goal of this policy is

> to improve livelihoods by mitigating and adapting to the adverse impacts of climate change, adopting a low-carbon emissions socio-economic development path and supporting and collaborating in the spirits of country's commitments to national and international agreements related to climate change.

A survey of Nepal's core climate policy documents reveals the development aspirations of the Nepali state's conception of climate change adaptation and the future it envisions, framed as green development and sustainable development (Sherpa 2021). However, economic prosperity is pursued in the context of climate change without a critical reflection on how that pursuit itself might be more destructive than the imminence of climate change effects. The systemic and structural injustice continue to benefit a few elites at the expense of the majority while pursuing national economic prosperity without IEJ concern. Even as resilient society and economy are envisioned, we see that various factors other than climate change that exacerbate risks and vulnerabilities at the local level are left out of consideration. Deep spiritual connections to the land and the environment that have sustained the places and its peoples in the Himalayas for generations, which are central to an IEJ framework are overlooked in institutional climate change adaptation (Figure 19.1).

Figure 19.1 Mountain cairn overlooking sacred Gokyo Tso (Lake) and Renjo La (Pass).
Photo credit: Author.

Institutional Adaptation and Everyday Adaptation

In January of 2020, news circulated about the Guinness World Record-making highest fashion show in Kalapatthar, close to the Everest basecamp. A Kathmandu-based clothing company had organized the show with models and other participants flown to the lap of Mount Everest. The organizers stated that their intention was to raise awareness about climate change. However, what that really meant – whose awareness was being raised and how using the Sherpa homeland as a backdrop to raise awareness – was not clear from the subsequent media coverage of this publicity stunt. The nuances of how climate change is unfolding in Khumbu and Pharak, and how the communities have been responding to climate change were absent in the media reportage. This was not an isolated occurrence. The framing of climate change in this and other similar events and initiatives exists in tension with an IEJ approach. A closer look at the numerous institutional climate change activities sheds light on this.

For two decades, governmental and nongovernmental institutions have organized climate change activities in Khumbu involving the Sherpa residents. Before that, institutional activities were framed around environmental conservation and development initiatives even though much of the engagement resembles what we consider climate change activities today. Institutional activities are distinct because they involve bureaucratic formalities such as the use of Nepali language to communicate and the documentation of the processes involved. An institutional activity begins and ends at a certain time with breaks in between on specific dates. During the activity, there are formal remarks to welcome the audience in the Nepali language with some occasional Sherpa and English. Minutes are kept and circulated with concerned authorities. In contrast, everyday climate change activities are determined by the decisions that the residents make for themselves. The realism of living with climate change for a yak herder might involve the decision to stop this livelihood practice due to restricted access to greener pastures higher up, and the opportunities in the financially lucrative tourism industry down valley.

If not carried out carefully, institutional climate change interventions in the region can also become sources of risks. In one instance, climate change researchers and development practitioners organized a potential glacial lake outburst flood (GLOF) research findings knowledge-sharing event in a popular tourist village in Khumbu (Sherpa 2015). The information did not reach the villagers as intended and instead became a source of rumor about an impending GLOF. A young mother with her newborn child was among many who had to find shelter in the middle of a rainy night to protect themselves from the rumored GLOF. Villagers in multiple places gathered their valuables and ran to higher grounds for safety in the dark. A villager who lives close to the critical Imja glacial lake was later quoted as saying that they would rather die once instead of having to fear dying again and again every time someone conducts scientific research in the region (Khadka 2012). Although one might be inclined to dismiss the incident as a random case of rumor, it deserves careful attention because of how that can jeopardize the safety of the villagers.

The assumption embedded in many institutional climate change adaptation efforts is the major fault in thinking that by reaching out to a few individuals in the region, the whole will benefit. Communities at the local level are not homogeneous units. Social heterogeneity even when working with a small-scale cultural world like that of the Sherpas from Khumbu should be understood as an influential and dynamic characteristic. It has been found that the same group of middle-aged, male hotel owners from on-route villages is generally targeted by the institutional climate change activities when the approach is not adjusted to fit the local networks and conditions. Simultaneously, such an approach overlooks the invaluable firsthand knowledge about climate change impact from farmers and herders of all genders. The marginalization of the farmers' and herders' knowledge as scientifically inadequate in addressing the seriousness of climate change does not help. It only further concretizes the knowledge hierarchy.

Institutional climate change adaptation activities largely aim to raise awareness of the residents about climate change and to enhance the capacity of the local residents to deal with climate change (Sherpa 2012, 2014, 2015). They tend to be short-term, techno-managerial, and active primarily along the main tourist trail to Mount Everest. Climate change in these activities is typically imagined institutionally: centering national, regional, and global scale climate change priorities and perspectives. Such a conception of climate change is then brought to the residents at the local level. This approach is evident in the narrow focus on glaciers and the melting of snow, which limits our understanding of the many ways the high mountain region and its peoples are living with climate change. While these climate change effects are not unimportant to local people, they reflect the priority of the national and regional scale politics of securitizing water resources. An IEJ approach to climate change adaptation reveals these discrepancies and allows us to recognize the plurality of responses. In 2011, Tengboche Rinpoche, the abbot of Tengboche Monastery placed *bumpas* (Sherpa for: sacred vessels) throughout Khumbu and Pharak as an offering to the local deities to protect the region from the harms of climate change. From a Sherpa perspective, the success of climate change adaptation requires not only assessing the increase in the volume of glacial lakes and disseminating that information to fellow residents, but also ensuring the reciprocal relationship with local deities and spirits of the land.

Concluding Remarks

Nepal's climate change policies in their existing form fail to recognize how Nepalis have already been adapting to climate change, or where support is needed (Sherpa 2021). Recent studies have revealed that biophysical, techno-managerial, and apolitical approaches to climate-change predominate in Nepali climate policies (Nagoda and Nightingale 2017; Ojha et al. 2015; Sherpa 2021). Biophysical understandings of climate change include predetermined adverse effects for different regions. In Khumbu, it is GLOFs. These policies consider human systems as separate from natural systems, and people are treated as apolitical passive recipients who primarily live in rural regions (Ojha et al. 2015). Climate change is also perceived solely as a static problem rather than a dynamic process thus limiting the understanding of its multidimensionality. Considering everyday lives of people at the local level, researchers have asked whether the institutional climate change response in Nepal is a 'missed opportunity' (Wong 2019) for governmental policies to address climate change effectively (Ensor et al. 2019), or if it represents an emergence of new discourses but the same old development approaches (Nagoda 2015; Nagoda and Eriksen 2014; Nagoda and Nightingale 2017) that fail to bring transformational changes to assist the most marginalized.

In extending this assessment with an IEJ lens, it becomes clear that if inclusion of populations like the Sherpas and other ethnic groups is to be promoted, as Nepal's climate-change policies claim to do, efforts would have to involve more understanding across epistemological differences instead of including selected cases of Indigenous knowledge. This would in turn require explicit discussion of what those differences are (Yeh 2016). Spotlighting Indigenous knowledge as important (Salick et al. 2020) without centering Indigenous peoples and their worldviews is not sufficient for IEJ. Before Indigenous knowledge can have any significant influence on climate change policy, however, the validity of Indigenous knowledge in general would require greater recognition at the national level (Tanner and Allouche 2011: 11).

The Himalayan glacier-fed rivers are frequently presented as the water source for more than a billion living downstream in the Indian sub-continent, therefore highlighting the need to investigate climate change impact in the broader region. While true, the critical oversight here is the many ways in which the mountains, the glaciers, and the rivers are also significant to the communities on the frontlines of climate change (Campbell 2017; Gagné 2020; Gergan 2015, 2017) in addition to being water resource to be managed for downstream urban mega centers such as Dhaka or Delhi. The reciprocity that maintains the relationships between humans and more-than-human beings is refused in the current mainstream climate change adaptation efforts (Chakraborty et al. 2021). An IEJ-deficient perspective on climate change adaptation in the Himalaya ignores the relationships between the residents and the places that have taken care of each other for countless generations. In contrast, an IEJ-informed climate change policy would be built on the principles of relationality, respect, and reciprocity without separating environmental concerns from spiritual ties to the land, mountain, air, and water, and with a special attention to the social, political, and economic survival of Indigenous communities.

References

Campbell, B. (2017). Encountering Climate Change: Dialogues of Human and Non-Human Relationships within Tamang Moral Ecology and Climate Policy Discourses. *European Bulletin of Himalayan Research*, 49, 59–87.

Chakraborty, R., Gergan, M. D., Sherpa, P. Y., & Rampini, C. (2021). A Plural Climate Studies Framework for the Himalayas. *Current Opinion in Environmental Sustainability*, 51(February), 42–54. https://doi.org/10.1016/j.cosust.2021.02.005

Ensor, J. E., Wennström, P., Bhatterai, A., Nightingale, A. J., Eriksen, S., & Sillmann, J. (2019). Asking the Right Questions in Adaptation Research and Practice: Seeing beyond Climate Impacts in Rural Nepal. *Environmental Science & Policy*, 94(January), 227–236. https://doi.org/10.1016/j.envsci.2019.01.013

Gagné, K. (2020). The Materiality of Ethics: Perspectives on Water and Reciprocity in a Himalayan Anthropocene. *Wiley Interdisciplinary Reviews: Water*, 7(4), e1444.

Gergan, M. D. (2015). Animating the Sacred, Sentient and Spiritual in Post-Humanist and Material Geographies. *Geography Compass*, 9(5), 262–275. https://doi.org/10.1111/gec3.12210

Gergan, M. D. (2017). Living with Earthquakes and Angry Deities at the Himalayan Borderlands. *Annals of the American Association of Geographers*, 107(2), 490–498. https://doi.org/10.1080/24694452.2016.1209103

Government of Nepal. (2011a). *Climate Change Policy*. Kathmandu: Government of Nepal.

Government of Nepal. (2011b). *National Framework in Local Adaptation Plans for Action (LAPA Manual)*. Kathmandu: Government of Nepal.

Government of Nepal. (2013). *An Approach to the Thirteenth Plan (FY 2013/14–2015/16)*. Kathmandu: Government of Nepal.

Helvetas and RRI. (2011). *Nepal's Climate Change Policies and Plans: Local Communities' Perspectives*. Environment and Climate Series (Environment and climate series 2011/1). Lalitpur: Helvetas Nepal and Rights and Resources Initiative.

ICIMOD. (2023). *Water, Ice, Society, and Ecosystems in the Hindu Kush Himalaya: An Outlook*. P. Wester, S. Chaaudhary, N. Chettri, M. Jackson, A. Maharjan, S. Nepal, & J. F. Steiner (Eds.). ICIMOD. https://doi.org/10.53055/ICIMOD.1028

IPCC. (2023). Climate Change 2023: Synthesis Report. A Report of the Intergovernmental Panel on Climate Change. Contribution of Working Groups I, II and III to the Sixth Assessment Report of the Intergovernmental Panel on Climate Change [Core Writing Team, H. Lee and J. Romero (eds.)]. Geneva, Switzerland: IPCC.

Khadka, N. (2012). Everest Sherpas in Glacial Lake Study Warning. Retrieved 11 April 2020 from https://www.bbc.com/news/science-environment-19569256

McGregor, D. (2021). Indigenous Environmental Justice: Towards an Ethical and Sustainable Future. In B. Hokowhitu, A. Moreton-Robinson, L. Smith, C. Anderson, & S. Larkin (Eds.), *Routledge Handbook of Critical Indigenous Studies* (pp. 405–419). London: Routledge.

Ministry of Environment. (2010). *National Adaptation Programme of Action to Climate Change*. Kathmandu: Government of Nepal.

Nagoda, S. (2015 November). New discourses but same old development approaches? Climate change adaptation policies, chronic food insecurity and development interventions in northwestern Nepal. *Global Environmental Change*, 35, 570–579.

Nagoda, S., & Eriksen, S. (2014). The Role of Local Power Relations in Household Vulnerability to Climate Change in Humla, Nepal. In T. H. Inderberg, S. Eriksen, K. O'Brien and L. Sygna (Eds.), *Climate Change Adaptation and Development: Transforming Paradigms and Practices* (pp. 200–218). New York: Routledge.

Nagoda, S., & Nightingale, A. J. (2017). Participation and Power in Climate Change Adaptation Policies: Vulnerability in Food Security Programs in Nepal. *World Development*, 100, 85–93.

National Planning Commission. (2011). Climate Resilient Planning: A Tool for Long-term Climate Adaptation. National Planning Commission. Retrieved 12 July 2018 from https://www.npc.gov.np/images/category/climate_resilent_planning.pdf

Ojha, H., et al. (2015). Policy Without Politics: Technocratic Control of Climate Change Adaptation Policy Making in Nepal. *Climate Policy*, 16(4), 415–433.

Salick, J., Staver, B., & Hart, R. (2020). Indigenous Knowledge and Dynamics among Himalayan Peoples, Vegetation, and Climate Change. In M. Welch-Devine, A. Sourdril, & B. Burke (Eds.), *Changing Climate, Changing Worlds: Local Knowledge and the Challenges of Social and Ecological Change* (pp. 55–69). Cham: Springer.

Sherpa, P. Y. (2012). *Sherpa Perceptions of Climate Change and Institutional Responses in the Everest Region of Nepal*. Dissertation. Washington State University.

Sherpa, P. Y. (2014). Climate Change, Perceptions, and Social Heterogeneity in Pharak, Mount Everest Region of Nepal. *Human Organization*, 73(2), 153–161.

Sherpa, P. Y. (2015). Institutional Climate Change Adaptation Efforts Among the Sherpas of the Mount Everest Region, Nepal. *Research in Economic Anthropology*, 35, 3–23.

Sherpa, P. Y. (2021). Nepal's Climate-Change Cultural World. In P. Sillitoe (Ed.), *The Anthroposcene of Weather and Climate: Ethnographic Contributions to the Climate Change Debate*. New York: Berghahn Books. (pp. 220–246).

Tanner, T., & Allouche, J. (2011). Towards a New Political Economy of Climate Change and Development. *IDS Bulletin*, 42(3), 1–14.

Wester, P., Mishra, A., Mukherji, A., & Shrestha, A. B. (2019). Summary of the Hindu Kush Himalaya Assessment Report. *The Hindu Kush Himalaya Assessment - Mountains, Climate Change, Sustainability and People*. Cham: Springer Nature Switzerland AG. http://doi.org/10.1007/978-3-319-92288-1

Wong, P. (2019). The Vulnerable State and Technical Fixes. *Himalaya-The Journal of the Association for Nepal and Himalayan Studies*, 39(2), 4–17.

Whyte, K. (2021). Against Crisis Epistemology. In B. Hokowhitu, A. Moreton-Robinson, L. Smith, C. Anderson, & S. Larkin. (Eds.) *Routledge Handbook of Critical Indigenous Studies* (pp. 52–64). London: Routledge.

Yeh, E. T. (2016). How Can Experience of Local Residents be "knowledge"? Challenges in Interdisciplinary Climate Change Research. *Area*, 48(1), 34–40.

20 Women, Water, and Weather

Kavre Villages Adapt to the Increasing Impacts of the Climate Crisis

Sonia Awale

Introduction

The climate crisis is a water crisis in the Himalayas. The mountains are warming approximately 0.7°C faster than the global average because of a phenomenon called Elevation Dependent Warming (Wester et al. 2019). Rains are becoming more erratic, groundwater is depleting and glaciers have shrunk, forming big new glacial lakes, which are especially at risk of bursting in a seismically active zone like the Himalayas. These problems bring devastating consequences for nearly 2 billion people living in Nepal and downstream from High Mountain Asia.

In the mountains, foothills, and plains of Nepal, increasingly erratic monsoons are leading to agriculture failure and in turn fueling the migration of people (Hoermann et al. 2011). There are already examples of internal displacement or mass migration due to the lack of water leading to reduced agricultural productivity in Paachthar, Terathum, Khotang, Ramechap, Dolakha, Chitwan, Tanahu, and Salyan districts, among others (Massey 2010).

This study focuses on the village of Phulbari in Kavre east of Kathmandu. Families in Phulbari and surrounding villages are selling off their land and moving to cities, the Tarai lowlands, or overseas to work. Every family has at least one member (often the men) either in Kathmandu or in the Persian Gulf or Malaysia. The drying of perennial springs, particularly after the 2015 earthquake, has meant that farmers dependent on rain-fed agriculture and small-scale dairy entrepreneurs have been unable to make a living in the village.

These challenges have disproportionately affected Dalits because of the location of their homesteads on south-facing slopes and added to the drudgery of women. They now have to walk further to fetch water for household use as well as for their livestock and the irrigation of their vegetable patches. But in some ways, it has also empowered them. In fact, outmigration has led to the feminisation of rural Nepal in general and agriculture in particular (Bose n.d.). Women here have devised local solutions to conserve water and transform farming practices in ways that are best suited to the land and the people. They have introduced a wide range of initiatives including rainwater harvesting, recharging a traditional network of ponds, drip irrigation, and organic farming.

DOI: 10.4324/9781003371175-26
This chapter has been made available under a CC-BY-NC-ND 4.0 license.

Nepal has always had either too much water or too little, and climate change has only made things worse. If the state had prioritised the agriculture sector which still accounts for one-fourth of Nepal's GDP and provided farmers with functioning irrigation systems, these families would have been more secure even in the face of the climate crisis. Instead, climate change has become an excuse for planners and leaders to blame for their inaction.

As with everything else, the poor, women, indigenous communities, ethnic minorities, and Dalits get the short end of the stick. Environmental justice in the context of Nepal will not be achieved unless the prevailing structural inequalities are addressed alongside the impacts of climate change. The state must also involve the people and communities most affected in decision-making.

Study Area and Methodology

This is a retrospective study, comparing the findings from 2023 to that of 2017 about how a community in Kavre's Phulbari has been adapting to depleting water resources. The focus population here is women. Findings are based on seven interviews, in-person and by phone, as well as group discussions. Watershed expert Madhukar Upadhyay was consulted during the study.

A Case Study from Kavre: Women, Water Crisis and Outmigration

It is bright, sunny, and unusually warm in late January. There is not a cloud in sight, and the Himalayan peaks to the north from Ganesh Himal to Gauri Shankar are even more majestic than usual.

The mountainsides, wrinkled with terraced farms, are abloom with yellow mustard plants. Kavre's Phulbari serves as a respite from the city for urban dwellers, just two hours' drive from Kathmandu but away from the hustle and bustle of the city and without its hazardous air pollution.

The panorama of the mountains is scenic but is a worrying sight -- even the high peaks are devoid of snow. It has not rained in nearly five months. The last time there was such a long winter drought, unprecedented wildfires blazed across the country. And if one looks closely, the mustard flowers are dry and frayed, and the soil on the terraces is desiccated and dusty.

"It doesn't rain as it used to, we don't get the drizzles that used to last for two weeks and irrigate our fields. There is no *Saune jhari* or *Maghe jhari* (rains in the months of July and January respectively) for that matter," says Nanu Ghatane who leads a women's group here in Buchakot, Phulbari-2, referring to summer and winter rains. "This year has been especially bad, we haven't had rain since *Asoj* (September) and it is making our water problem even more severe."

Now, when it does rain, it pours. Such big rains can spur flooding and landslides. Too much water at once can also destroy the crops and vegetable plantations and does not help with groundwater recharge as most of the water runs off. Villagers have not had a good yield of rice, maize, soybean, and seasonal vegetables in recent years.

Figure 20.1 Laxmi Parajuli spends US$ 23 a month to buy water from a tanker for her herd of a few goats and cows.

Photo credit: Author.

Meanwhile, perennial springs have dried. In the last 10 years, 30 of the 40 springs in the village have gone dry while others have been reduced to a trickle, a problem made worse by erratic rain. Ghatane and four other families now rely on just one spring, *Patal ko dharo*. Even springs have names here because they are so important in the dry season from December to April. A couple of years ago, each family used to get ten *gagri* of water a day during winter. They now barely get enough to fill five of the traditional water pots.

This has meant that to meet their needs, families have had to buy water from private water tankers. The only other alternative is groundwater extraction but most families do not have enough money to afford wells and pumps.

Laxmi Parajuli's family has to bring water from the only working spring in the village. But that does not meet all her needs. She spends an additional Rs 3,000 a month (about US$23) to buy water from a tanker that ferries it from Rosi Khola, one hour drive away, so that her cattle, a few goats and two cows, have enough water. "If there was enough water I would have wanted to rear five cows," she adds (Figure 20.1).

Ghatane herself wanted to start poultry farming on the half-hectare of land she owns in the village. But she has not been able to because there is no water. There used to be a spring in her field but it dried up years ago. She also thought about planting avocados and even visited the agriculture division in the village. But the first thing they asked her was if she had enough water to sustain new plantings.

"My husband is in Kathmandu, he doesn't have much work there and what he earns is spent on rent and food. He wanted to come back and start up a business," says Ghatane. "But you need water before you start anything, whether it's poultry, vegetable farming, or horticulture. So he is unlikely to return anytime soon. My children are also unlikely to come back."

Krishna Pariyar, a Dalit man, lives alone with his wife at his ancestral home in Kavre. He has a family of 17 made up of children and grandchildren, but most of them live in the city. The couple had once traveled to Kathmandu for a week but did not like it and came back. "Is that even a place to live in? The food, the water, the pollution, I will never go back," says the 69-year-old. "For us, there is nowhere else like our village, it has immense potential."

But to harness that potential, the limiting factor for villagers is the lack of water. In fact, one of Pariyar's sons Dipendra had recently come home from Kathmandu with big plans to start a fishery and a pig farm. He tried everything he could but there was just not enough water to launch his venture. He has now gone back to the city.

The villagers have now pooled some money to pump the water up from the river down the mountain and have also received some financial support from the Ward office. Even then, the water will be just enough for household chores and a bit of vegetable farming for their families.

Buchakot in Phulbari-2 is an old settlement of 95 households (35 Tamang, 30 Brahmin and Chettri, and 30 Dalit). Many of them are dairy farmers. Kavre, like many of Nepal's mountainous districts, is depopulating because of young people moving away for jobs and opportunities in the city and abroad. The lack of water has accelerated this trend.

The district is well known for dairy and vegetable farms that used to supply milk and produce to nearby Kathmandu. It is where organic farming started in a big way in Nepal as farmers were encouraged to reduce the use of pesticides and chemical fertiliser.

But the lack of water has hit farming hard, and the only option for many is to move out. Five families, three Dalit and two Brahmin have left in 2023, with neighbours saying many would return if the water situation improved.

Besides women, it is the Dalit families who have been particularly badly hit. They tend to live in the drier south and east-facing slopes, where the springs went dry first. In previous decades, Dalits could only use their own taps and could not get water from springs used by the 'upper' castes.

Education, awareness, and laws against caste discrimination have mitigated this trend. But now, it is climate change that has limited water supply. "We are no more subjected to the kind of discrimination we used to face in the past, people are now educated and more aware," says Krishna Pariyar's daughter Gita who lives with her husband in a bazaar in Namo Buddha. "The bigger challenge today is sustaining our livelihood solely based on farm and cattle."

Surbir Biswakarma, a Dalit man, is carrying a *gagro* of water in a straw basket with a strap around his head up the steep hill, when he stops to catch his breath. He says he makes two to three trips a day from his home to the only functioning spring in the wetter and forested northern slope on the other side of the ridge. It takes an hour each roundtrip.

"Dalits have left in droves, they have sold their land and migrated elsewhere with better access to water and livelihood. Some Brahmin and Chhetri families have too," adds Ghatane.

Many of the Tamang population here have not moved away, though, but they have at least one family member either in the city or abroad to help them out with expenses. They have also moved to opening small shops instead of running their farms.

Govinda Parajuli had a thriving goat business in Phulbari but he had to buy a lot of water to keep it afloat. Eventually, the cost of water was too high for the firm to be feasible. His only alternative is an overseas job.

During the Covid-19 pandemic, many of the men who had migrated elsewhere did return but, as soon as conditions allowed, almost every one of them decided to go back to the Persian Gulf or Malaysia. Lack of water has meant that they cannot start any business in the village. Laxmi Parajuli's husband Maheshwor is a returnee migrant worker from Malaysia who doesn't seem confident about staying back.

"Our men toil in the desert heat for a mere Rs 30,000 a month. If only we had enough water, they can easily make the same money here, or more," laments Ghatane.

The vegetation of Kavre is such that it accommodates different kinds of crops, fruits, and vegetables, even ones mostly found in the warmer temperatures of the Tarai southern plains like mango, lychee, and pineapple. There is a dairy collection counter just half an hour from the village. Farmers get about Rs80 for every liter of cow milk and Rs140 for buffalo milk. On average, families produce anywhere between 5 and 10 liters, but they could do more if there was enough water.

Instead, the men are continuing to migrate overseas for work, adding to the burden of women who now head households, farms, and local businesses and serve on committees for community forestry, school management, and village drinking water supply, in turn empowering them.

Why Are Nepal's Springs Going Dry?

Watershed expert Madhukar Upadhyay has noticed springs going dry across the country since the 1990s. In particular, he has been monitoring the state of Jhiku Khola in nearby Panchkhal Valley. This stream at an elevation of 1,006 meters in the Kavre district is one of the most extensively used watersheds in the foothills of Nepal. It used to overflow during monsoon and supply water all year round but has been seeing a gradual decline.

In previous years, Panchkhal itself had a thriving economy, supplying 60% of Kathmandu's daily vegetable demand. The fertile landscape and its proximity to the capital meant that organisations invested in farmers, especially women, and soon the region was a centre of agriculture, vegetable farming, and horticulture. But this also meant increased demand for water and before long the water resources were exhausted.

"This is the first time I have been to Jhiku Khola since I visited back in 2019. We were already losing it back then but this time its decline has reached a new height," Upadhyay noted in 2023. "Jhiku is now a dumping site for nearby dairy and a swine farm. Only the springs that are on the lower belt have some water left."

Springs are going dry across Nepal's high mountains and middle mountains. Even parts of the southern plains in the Tarai have noted wells going dry because of the lowered water table. In fact, researchers say the whole of the Himalayan belt is experiencing an unprecedented water crisis.

A study published in *Water Policy*, the journal of the World Water Council Adhikari et al. (2020), mapped over 4,222 springs from five different watersheds in the western mountains of Nepal. Based on the estimation of discharge flow in these springs, the paper argues that about 70% were seeing a decline over the last ten years.

The Nepal Planning Commission (NPC) in 2013 conducted a study covering Mahottari, Udaypur, Paachthar, Kavre, and Gulmi districts on how their environmental vulnerabilities including flooding, landslides, and droughts were leading to the displacement of the locals and in turn fueling outmigration. It found that water sources in Panther and Kavre have been drying up over the previous decade while in Mahottari, women who are tasked with household chores have been hit the hardest due to the declining groundwater table.

Not surprisingly, the paper states 'farmers reported sharp declines in the production of rice, maize, millet, wheat, mustard, peanuts, buckwheat, broom grass, ginger, lemons, and oranges.' In parts of Panchthar and Gulmi, people have stopped planting rice because conditions are too dry. 'Practising animal husbandry has become difficult, and the production of milk and meat has declined in drought-hit areas.'

The 2013 paper also found that while people from Panchthar and Gulmi have coped with water shortages by moving to Tarai districts, local Kavre residents have moved from highlands to lower elevations where more water is available with hopes to return to higher ground once the water situation improves.

Nepalis have often migrated. In the 1960s, they moved from the hills to the Tarai after the control of malaria and later sought better economic opportunities, healthcare, education, and infrastructure down south. In the late 1990s, it was the Maoist conflict that drove Nepalis to migrate, especially from the western regions to Tarai towns and urban centres. Consequently, 53% of Nepal's 30 million population now live in the southern plains.

But now people are increasingly migrating because of the lack of water. And one of the primary reasons for the prolonged drought is thought to be climate change.

The Central Bureau of Statistics' 2016 Climate Change Impact Survey interviewed Nepalis across the country. Seventy-five percent of respondents said that they have observed changes in their water resource of which 84% reported a decrease in the amount of surface water. Likewise, 75% of the households in the high mountains have noted complete dying of surface water while nearly 50% in the middle mountains have noted drying of the underground water sources. 'Such changes in water sources have been reported due to insufficient rainfall,' states the report.

Researchers put the blame on climate change. 'Warmer temperatures enhance evaporation, which reduces surface water and dries out soils and vegetation. This makes periods with low precipitation drier than they would be in cooler conditions,' states the US-based Center for Climate and Energy Solutions (Author Year). 'Climate change is also altering the timing of water availability. Warmer winter temperatures are causing less precipitation to fall as snow... Decreased snowpack can be a problem, even if the total annual precipitation remains the same. This is because many water management systems rely on spring snowpack melt.'

Climate change is even more pronounced in the Himalayas because it is warming much faster than the global average. But having said that, the drying of the springs cannot be blamed only on climate change. Changes in land use, deforestation, and urbanisation all play a role. Haphazard planning and bad engineering in particular is a big culprit. The concrete house-building spree across Nepal, even in villages, has decimated much of our forests and water sources.

Villagers notice the changes. "We are not scientists, we can't say why the springs are drying up but even we know that we haven't been able to protect them," says Nanu Ghatane in Kavre. Ghatane continues,

> There used to be a big stream right below us called Bas Pokhari. It was so large and secluded that people avoided it even during the daytime. They thought it was haunted. It recharged all of our lands but then we dug up a road and the water dried up.

Kavre is now seeing an increasing trend of reverse migration, wherein people and businesses from the cities are buying plots of land and entire terraces in the mid-hills to build resorts and hotels. As the demand for water shoots up, businesses will extract groundwater, further depleting already scarce water sources. In the meantime, as villages build up and turn into concrete jungles, natural groundwater recharge is reduced.

"We just pump out, we do not replenish groundwater," says Upadhyay. "Even so, over-extraction is a major issue in the Tarai but it should not have led to the drying of the spring in the hills."

One could argue that the replenished community forest near Phulbari should recharge groundwater sufficiently as well as keep the springs from drying up. But that has not been the case, unfortunately. Nepal's community forestry programme is widely recognised internationally for doubling the country's tree cover in 25 years but a largely ignored aspect of this success is the poor quality of these forests and their management (Ojha 2023).

The 2015 Nepal earthquake is often also cited by researchers and locals alike as a cause leading to the large-scale drying up of the springs in the country's mid-mountains. In Kavre, too, villagers say that many springs dried up after the earthquake. Water experts also do not discount this link calling it a natural outcome given that the 7.8M and 7.3M tremors shifted the ground both horizontally and vertically.

Shifts in the mountains, they say, led to changes in the water flow channels underground where people have noted springs going dry, water flowing from springs that were non-functional, or new springs emerging. Meanwhile, according to the government, over 5,000 springs have dried up or been destroyed in the 14 districts most affected by the earthquake.

Some studies corroborate the idea (Khanal 2016). Barfal et al., (2022) studied the 2011 6.9M Sikkim earthquake and the resulting impact on the water springs. 'Crustal deformation, shaking, and movement of the earth's surface due to an earthquake, can modify the stream flow and water level in wells through consolidation of surficial deposits and development of new fractures,' states the paper. It further noted: 'Our study suggests that in tectonically active mountain ranges, like Himalaya, an earthquake not only causes surficial deformations but also influences the hydrological framework at a regional scale. Therefore, the seismic nature of the terrain should also be considered for spring rejuvenation policies (Barfal et al., 2022: 1).'

Local Initiatives and Solutions

Increasingly worried about depleting water resources and its impact on their livelihood, in the early 2010s Nanu Ghatane with support from other women in Phulbari village set out to find the answer to their problems. The team of citizen scientists formed various groups including senior citizens, women, and youth. Ghatane would hold meetings with each of these groups frequently and ask them what their biggest problem was.

Water shortage was at the top of the list for every group, followed by reduced agricultural productivity. The elderly in particular were consulted about the rainfall patterns in the past and historical droughts. They also had valuable knowledge about the decreasing moisture content in the soil and the adverse effects of using chemical fertilisers.

After identifying their problems, Ghatane wrote applications detailing them to local government offices who then sent her to attend a week-long workshop on soil conservation and reviving water springs organised by International Centre for Integrated Mountain Development (ICIMOD) in Melamchi in 2014.

Upon her return, Ghatane formed *Samudayik Batawaran Samrachyan tatha Jalawayu Anukulan Samuha* (Community Environment Protection and Climate Change Adaptation Group) with 95 households as members. One of their first undertakings was to identify and map water springs in their locality as well as their status. They found that 30 of the 40 springs had gone completely dry in the previous 10 years.

The next plan of action was to protect the springs that were still functioning while also trying to revive the rest by building ponds. At the household level, villagers installed a rainwater harvesting system such that the monsoon runoff is collected in *Ghampe* tanks of varying capacities. Many built plastic ponds to store wastewater to be reused for their vegetable patch via drip irrigation.

But the 2015 earthquake struck and their work on reviving the ponds had to be stopped in favour of rebuilding the houses that were damaged or destroyed during the disaster. A few years down the line everyone was hit by the Covid-19 pandemic which further delayed their plans.

But during this time they found out that plastic ponds are impractical. They do not recharge groundwater and they are also prone to wear and tear. Drip irrigation was also not as effective in their topography and led to unequal distribution of water even in the same vegetable patches.

Ghatane and her group of women are now getting back to reviving the ponds and protecting those that are still functioning. They are also reforesting their forests with locally native tree species.

Conclusion: The Way Forward

There is no one reason for the drying up of springs in the Himalaya. Climate change, changing land use, deforestation, earthquake, and over-extraction of groundwater all play a role. Depleting water resources is adding to the burden on women while men have migrated, seeking economic opportunities elsewhere. It must be said that the water crisis is proof of structural inequalities, a lack of empathy for the people at the margins and with limited resources and political mismanagement and inaction.

When addressing the impacts of climate change, Nepal's planners must also take into account the prevailing issues of gender, class, and caste. It is women and Dalits who are hit in the greatest numbers by Nepal's changing waterscape. It is imperative that the people most affected by the climate crisis are capacitated to adapt. Above all, the water crisis can only be addressed fairly when there is a true devolution of power with the central government backing local communities.

In the meantime, communities have to take the lead much like in this village in Kavre. And while external organisations can bring in expert opinion and technical support, pilot projects are too small in scale to make much difference. If the interventions are working, they must be scaled up, so that other communities can learn and adapt better.

"The only way to revive the springs is to build ponds and this should be done as a voluntary campaign and not as a project because a piecemeal approach is not going to be sustainable," says Madhukar Upadhyay. Upadhyay continued,

> I say we deploy our high school students across the mid-hills to dig ponds, small table-sized ponds in close proximity throughout watersheds. And the timing has to be right, it has to be done much before the onset of monsoon.

But most of all, Nepal needs to study the science and hydrology of its water sources, recognise the depleting water sources as the crisis it is, and formulate and implement policies that take both the people and ecology into account.

References

Adhikari, S., Gurung, A., Chauhan, R., Rijal, D., Dongol, B. S., Aryal, D., & Talchabhadel, R. (2020). Status of Springs in Mountain Watershed of Western Nepal. *Water Policy*, *23*(1), 142–156. https://doi.org/10.2166/wp.2020.187

Barfal, S. S., Das, M. M., Joshi, M., Joshi, R., Kumar, K., Kumar, D., & Rai, Y. K. (2022). Response of Water Springs towards an Earthquake: A Case Study from Sikkim Himalaya. *Journal of Applied Geophysics*, *206*, 104792. https://doi.org/10.1016/j.jappgeo.2022.104792

Bose, I. (n.d.). Migration in Nepal through the Lens of Climate Change Dhaka: Climate Action Network South Asia.

Hoermann, B., Banerjee, S., & Gerlitz, J. Y. (2011). *Labour Migration as a Response Strategy to Water Hazards in the Hindu Kush-Himalayas*. https://doi.org/10.53055/ICIMOD.546

Khanal, K. (2016, April 25). *Water Sources Run Dry in Nepal After 2015 Earthquake, Forcing People to Leave Their Ancestral Villages or Hike to Faraway Wells*. Global Press Journal. Retrieved from https://globalpressjournal.com/asia/nepal/water-sources-run-dry-nepal-2015-earthquake-forcing-people-leave-ancestral-villages-hike-faraway-wells/

Massey, D. S., Axinn, W. G., & Ghimire, D. J. (2010). Environmental Change and Outmigration: Evidence from Nepal. *Population and Environment*, *32*(2), 109–136. https://doi.org/10.1007/s11111-010-0119-8

Ojha, H. (2023, January 3). *Reviving Nepal's Community Forestry Success Story*. Retrieved from https://nepalitimes.com/news/reviving-nepals-community-forestry-success-story

Wester, P., Mishra, A., Mukherji, A., & Shrestha, A. B. (Eds.). (2019). *The Hindu Kush Himalaya Assessment: Mountains, Climate Change, Sustainability and People*. Springer International Publishing. https://doi.org/10.1007/978-3-319-92288-1

21 Applying a Climate Justice Framework to Understand Inequities in Urban Water Governance Amid Climate Change Challenges in Nepal

Gyanu Maskey, Poshendra Satyal, Monika Giri, and Prajal Pradhan

Introduction

Climate change already affects regions worldwide with increased intensity, frequency, co-occurrence, and persistence of extreme events (Pradhan et al., 2022). Nepal is at high risk of climate change due to its fragile ecosystem, unstable geology, and complex topography (Gerlitz et al., 2015; MoHA, 2013). Considered one of the most vulnerable countries to climate change, Nepal is warming at the rate of 1.2°C per decade (Gautam et al., 2013; Mc Sweeney & Lizcano, 2008). In the recent past, the country has witnessed erratic rainfall, increase in mean temperatures, and unpredictable onset of monsoon seasons (MoE, 2010) with uncertain precipitation (Duncan et al., 2013; Poudel & Shaw, 2016), and more irregular and prolonged droughts (Poudel et al., 2019). These trends of climate variability and change have increased extreme events in the country in terms of flash floods, landslides, glacial lakes outburst floods, and droughts (Poudel et al., 2019), with severe impacts on the overall socio-ecological systems of the high mountains, mid-hills as well as plain areas of the country (Shrestha & Aryal, 2011).

For developing countries like Nepal, the impacts of climate change are even more complicated due to pre-existing social inequities. Several studies have illustrated that climate impacts are more severe for the poor and socially vulnerable groups than others (Islam & Winkel, 2017; Marino & Ribot, 2012; Thomas et al., 2019), compounded by the interaction with pre-existing inequities in caste, class, ethnicity, religion, gender, culture and spatiality. Such disproportional impacts have also been observed during recent disasters such as April 2015 Nepal earthquake (Shrestha et al., 2019) and the COVID-19 pandemic (Pradhan et al., 2021). Thus, climate change brings a new dimension to Nepal's evolving social and climate justice discourse (Satyal, 2011; Satyal et al., 2020).

There is widespread recognition that climate change adaptation must concurrently address the root causes of vulnerability, attempt societal transformation, and achieve climate justice (O'Brien, 2016; Pelling, 2010; Ziervogel et al., 2016).

DOI: 10.4324/9781003371175-27
This chapter has been made available under a CC-BY-NC-ND 4.0 license.

Some interventions for climate adaptation often inadvertently reinforce, redistribute, or create new sources of vulnerability (Eriksen et al., 2021; Satyal et al., 2021). These adaptation trajectories trap the poor and marginalized within vicious cycles (Henrique & Tschakert, 2021). Therefore, climate justice scholars and advocates emphasize procedural, distributional, and recognitive aspects. The shift needs to be towards transformative climate justice (Newell et al., 2021), interactional and mobility justice (Radonic & Zuniga-Teran, 2023), and emancipatory urban climate justice (Porter et al., 2020).

Still, we find limited literature on the differential impacts of climate change on various social groups. The discussion around human and social systems lacks sufficient analysis. Climate justice is globally well-established in philosophical and academic work and social movements. However, in Nepal, there is a limited understanding of how the meaning and discourse of climate justice are operationalized. Further, in Nepal, there has been a shortage of empirical studies on the differential impacts and responses of different social groups.

We address the above-highlighted knowledge gaps based on empirical case studies of two Himalayan towns in Nepal -- Dhulikhel and Diktel, focusing on equity and justice in water governance. Our chapter answers two main questions: (a) What are the key issues in the climate justice debate, and how can a climate justice framework be applied to analyze our case studies? (b) Which dimension of climate justice gets priority and more traction than others?

Methodology

Our qualitative analysis is based on primary and secondary data from the two urban water governance case studies. We conducted key informant interviews and focus group discussions with stakeholders including local communities differentiated across class, caste, class, ethnicity, gender, and spatiality, as well as actors engaged in developing and implementing water-related projects in these towns. Interviewees were members of the water management committee (women and marginalized members), communal taps, and project water users in the study areas. We develop these case studies based on our long-term engagement in water governance research in these sites and also draw our analysis from participant observation, transect walks, and participation in events. Further, we review existing literature related to water governance and climate justice in Nepal.

Dhulikhel and Diktel, the study sites, are both mid-hill hilltop towns facing growing water needs. Dhulikhel is located in east-central Nepal, 30 km from Kathmandu,[1] while Diktel is in eastern Nepal. Dhulikhel is a rapidly emerging satellite city of Kathmandu bearing prime importance from a commercial point of view, with a population of 66,405. The water supply is 13.8 MLD (Million Litres per Day), whereas the demand is 23.1 MLD. The town taps water from a distant Roshi River 14 kilometres away. Diktel with a population of 46,903, benefits from numerous spring water sources, rivers, and small streams in the municipality. However, with the increasing population, rising commercial and tourism activities, the demand of water has grown significantly in the town.

Conceptual Framework for Thinking about Climate Justice

We develop and adopt a climate justice framework for our analysis. It considers distributional, procedural, and recognition issues and aspects of vulnerability, resilience, and adaptive capacity (Berrang-Ford et al., 2021; Sikor et al., 2013; Walker, 2012) (Table 21.1).

Distributional justice is allocating rights and responsibilities or benefits and burdens fairly among various actors. It is about identifying who gets what and how much, who wins and losses from climate impacts and policy interventions directed to address climate impacts. 'Procedural justice' is about identifying who is involved in decision-making, how various actors are involved, what procedures apply the decision-making, and what institutions oversee it. *Recognition* is about knowing who is recognized as a legitimate actor and how their concerns, needs, and interests are represented (Figure 21.1).[2]

Other relevant aspects of climate justice include intergenerational, relational, and transformative dimensions (Gulliver et al., 2023). The intergenerational aspect represents the extension of climate change's causes and impacts linking communities temporally. The cooperative relationships among groups purposely aiming to advance climate justice are the relational aspect. The transformative aspect is the remaking of power structures rooted in social and institutional inequalities driving and perpetuating climate change and responses to it.

Results and Analysis

We first discuss key issues in the debate around climate justice in Nepal and then proceed to analysis of urban water governance case from justice lens.

Table 21.1 Three dimensions of justice and their application in the case studies

Dimension of justice	*Definition in a climate change context*	*Urban water governance case*
Distributional Justice	Encompasses the distribution of impacts and vulnerabilities and the distribution of assistance to adapt to these impacts (Adger et al., 2006)	Spatial inequity (core-periphery divide and topographical differences) resulting in inequitable water distribution (and quality)
Procedural Justice	The ability to participate equally in fair institutional and decision-making processes (Schlosberg, 2007)	Dominant male participation, core community members in decision-making and water forums (participatory exclusion), women engaged in minor decision-making
Recognition	Issues of injustice relating to the lack of recognition of individuals or cultures, including degradation, devaluation, or oppression (Honneth, 2001; Schlosberg, 2007)	Upstream peoples' local water rights claims to prior rights versus downstream town dwellers' claims to water use being in the same municipal jurisdiction

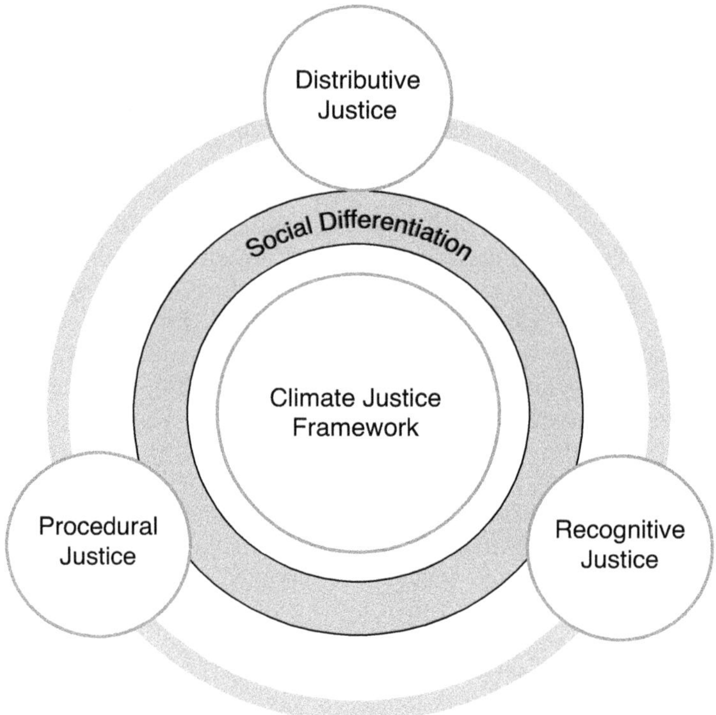

Figure 21.1 Climate justice framework: social differentiation and key dimensions of justice.
Source: The authors.

Key Issues in the Debate on Climate Justice in Nepal

Studies in Nepal reveal various forms of differential impacts of, and responses to, natural hazards and climate change, most notably for socially marginalized groups (Bhattarai et al., 2015; Gentle et al., 2014; Macchi et al., 2015). The country's inequitable agrarian political economy suggests disproportionate climate change impacts across groups. Furthermore, responses to natural hazards often fail to prioritize the poorest and most vulnerable groups. Inherent social differences are a problem. The unequal social structure of Nepal is divided along lines of caste, class, ethnicity, religion/culture, region, language, gender, and age.

Still, the literature on climate justice concerns is relatively new in Nepal. Few studies on the environment and natural resources address environmental and climate justice issues. For example, studies have highlighted politics of knowledge and stories of climate-society relationships (Chakraborty & Sherpa, 2021). However, discussions have yet to be transformed from climate adaptation, resilience, and mitigation to climate justice issues (Chakraborty & Sherpa, 2021). Even debate about environmental justice is scarce in Nepal.

The idea of just transition and climate justice has also been studied through the lens of food sovereignty and social movements (Routledge et al., 2018). Some studies focus on how carbon financing is used for greater climate justice in Nepal (Manandhar & Bhatta, 2013). Furthermore, climate justice issues have also been linked with inequitable access to climate information and technology (Homberg & McQuistan, 2019) and allocating resources and funding to those experiencing climate change impacts (Homberg & McQuistan, 2019; Madhanagopal et al., 2022)

Case Studies of Urban Water Governance in Dhulikhel and Diktel

Our urban water governance case studies in Dhulikhel and Diktel highlight how adaptation interventions aimed at inclusive and sustainable access to water supply services marginalize the poorer sections of society. In both sites, the Asian Development Bank (ADB) has funded projects to address increasing water insecurity amid the changing climate. The ADB project explicitly targets improved water supply and sanitation infrastructure and strengthened institutional and community capacities. However, amid climate risks in Dhulikhel and Diktel, urban water governance faces equity and justice concerns. These concerns are related to the distribution of water services and decision-making processes. In addition, claims and contestations around access to water sources have been documented, raising concerns about distributive, recognitive, and procedural justice.

Distributive Justice

Our analysis highlights that among residents in both study sites, the water distribution is inequitable. The poorer sections of society (Dalits, women, and ethnic groups) face unequal access to piped water supply owing to spatial inequity and they are not able to afford measures to address it.

Our cases highlight the spatial inequity of water distribution between the core and periphery areas. Water supply schemes focus on distributing water to the core areas in Diktel and Dhulikhel, often home to economically well-off people, while people in peripheral areas are deprived of piped water. This problem is due to the municipality annexing nearby peri-urban and rural areas beyond the planning of large-scale schemes. In Dhulikhel, the annexation of the nearby villages in 1986, 2015, and 2017 has led to the exclusion of the newly added wards to water supplied by the Dhulikhel Drinking Water and Sanitation Users Committee (DDWSUC). DDWSUC provides water only to core wards 3, 4, 5, 6, 7, and 8, but not to periphery wards 1, 2, 9, 10, 11, and 12. An ADB-funded drinking water project, Kavre Valley Integrated Water Supply Project (KVIWSP), near completion in Dhulikhel, Banepa, and Panauti, has also not yet targeted these peripheral areas for water service.

Likewise, in Diktel, the ADB-funded drinking water scheme has initiated water distribution only to wards 1, 2 and 3 of Diktel Rupakot Majhuwagadi Municipality.[3] The geographical coverage of the ABD scheme in Diktel addresses the area previously covered by two projects: the *Gaun Khane Pani* (village drinking water) and *Shahar Khane Pani* (town drinking water), which have merged to form a single committee. The user committee claims that the project covers the major water-scarce areas of Diktel, indicating both the core bazaar area and peripheral areas.

The varied topography, ranging from flat land to hilly areas within the same ward, also creates inequitable water supply (Maskey et al., 2021). Poor people in the hilly areas face difficulty accessing water owing to low water pressure; they cannot afford water pumps to pull more water. In Dhulikhel, the drinking water users committee has restricted the use of machines to pump water.

Further, the new large-scale water scheme in Diktel has removed communal taps,[4] with mandatory individual connections. Poorer communities cannot afford the installation charge (NPR 65,000, increased from NPR 35,000), and the monthly tariff. Those communities previously used water from the community taps and divided the tariff (Rs 200) among the households. Hence, they now completely lack water services. Contending that the charge is minimal, the water user committee members showed no concern for those who might face a problem.

Inequities in water access based on caste and gender dimensions are also noted in both Dhulikhel and Diktel. Marginalized communities such as Dalits have often struggled to access water. Now the usual sources where they used to fetch water have started drying up due to changing climate. On top of this, large-scale water supply schemes have further marginalized their water access. In our focus group discussion, Dalit men revealed that they were longing for water access but did not know where to go and whom to ask. In contrast, water users committee members mentioned that some of the 47 remaining connections are going unused. Some of them had to be reallocated for "*mananiya jiu* (local politicians), family and *thula manche* (local elites)."

Our interviews also highlighted that household water responsibilities were entirely shouldered by women. At a focus group meeting with women in a village in Diktel municipality, one woman in her fifties said: "Women are more efficient in water use and use only about half of water as men would use for the same purpose, i.e., cleaning dishes and cloths." This sounds like strategic household water management but it adds a burden to women.

In some cases, gender exclusions and caste exclusions combine, even despite progressive changes in the last few years in Nepal. For instance, in Dhulikhel women from higher caste groups were reluctant to allow women from marginalized Dalit communities to touch the spring sources, and even resting their water pots in the stone was not allowed (Shrestha et al., 2022). Likewise, conflicts between Dalits and non-Dalits over the use of public tap water have been noted in Dhulikhel (Pokharel et al., 2018). In Diktel, conflicts between women, especially from the Dalit caste, were reported. They had to compete fetching water early morning from the limited water source and the situation is getting worse as the water sources are drying.

Procedural Justice

We find that control over water-related decisions is limited because of pre-existing social differences. In both towns, water user committees tend to include representatives from the core area who are relatively powerful and wealthy households. For instance, in Dhulikhel, the elite control the water security systems with no participation of people from the annexed or the peripheral areas. This was observed since the founding of the community in 1992. Here, the Newars of the core areas, an ethnic group mainly engaged in the business sector, dominated the committee (Pokharel et al., 2018). In the Drinking Water Supply Management Board Act 2006, including women is a priority but not mandatory. Women are likely to be excluded from being a board member; none of the Water Board organizations are headed by women.

In Diktel, we were told that the committee for the ADB project formed by merging the committees of two rivers (*Ramba khola* and *Majh khola*) was formed with representatives of the political parties. One woman member of Diktel told us about her struggle for more female representation. According to the rules, committees with a male chair must have a female vice-chair, but the provision has not been followed. There are no female members in the 13-membered Advisory Committee of the ADB project.

In Dhulikhel although some women have started having their voices heard in the water-related committees, others, particularly from low socio-economic backgrounds, single women and Dalits, still find it challenging to participate in decision-making structures and processes. A woman member of Diktel's water users committee shared her happiness in having a say in choosing the colour of the paint of the committee building. However, male members controlled most major decisions, such as on fund utilization. We also found that in Dhulikhel women members of the DDWSUC were offered limited roles, for example, organizing cleanliness campaigns in the community. Their voices of opposition were neglected regarding making changes in the water tariff and deciding the location for new deep well boring. The woman vice-chair also shared that the male members reversed decisions in the meetings as per their convenience.[5] These testimonies show that stereotypically 'feminine' roles were offered to women while male members were offered technical and higher-level decision-making roles. Moreover, we learnt from the social mobilisers and witnessed in Diktel during our field visit that the committee meeting was usually in the morning, which prevented or reduced participation by the women members owing to their household responsibilities.

Recently, in Dhulikhel the DDWSUC has considered including women representatives on the committee, i.e., six female and nine male members. However, the increased number of women on the committee does not mean all the members will speak out in the meetings. As reported, the majority of the women members (four out of six) said they do not speak against the non-participatory decisions as they are not aware of their rights or do not want to challenge the status quo. Anyone who speaks out faces difficulties securing their position on the committee.

Related to procedural justice, in the 'water forums' (*Paani Chautari* in Nepali),[6] although a few females were attending, none except the female Deputy Mayor raised and discussed any issues, including issues specifically affecting women. Females were not adequately engaged in the discussion although the facilitators tried to create an inclusive environment for presenting views. This problem highlights the need to give trainings on leadership enhancement and provide capacity building for effective participation and empowerment of women.

Recognitive Justice

In both study cases, rights contestations between the upstream and downstream communities were very salient. In such contestations, claims of justice and injustice often featured recognition issues, for example, who can be recognized as a 'local user'. In one of the discussions in Diktel, the secretary of the water users committee argued strongly for the upstream people claiming their rights to natural resources, including water sources, even referring to the provisions in the International Labour Organisation Article 169 (ILO-169). They claimed their inheritance to the water sources from their forefathers, which they hold as their fundamental rights. In contrast, the people residing in the core part of the town downstream argue that they also live in the municipality and thus deserve water rights. We find the claims of rights by the upstream people, where the source is located, to be unfair. In addition, there are also instances where the upstream households have not paid the full tap installation charge, arguing for their rights to the water sources of their area: One female interviewee declared, "Why should we pay such huge amount for our own water?"

The committee members believe that such claims are illogical. The amount of water cannot be more just because it belongs to the area where the water supply is located. According to the committee members, "*Jal Jungle ra Jamin* (water, forest, and land) – country's natural resources belong to the government, with equal rights to all." In Dhulikhel's case, the upstream communities argue that they have prior rights as they depend on the same water for irrigation and water mills operation. They do not have excess water to offer to the town dwellers. A major reason for increasing contestations is unmet expectations, as water-related projects did not fulfil the commitments made to the upstream users during the start of the project.

Comparing the Three Types of Justice

Among the three key justice dimensions, in both Diktel and Dhulikhel, distributive and procedural concerns are most apparent; recognitive justice was the least apparent. The issue of water scarcity has been evident due to the increasing urban populations and climate change problems. For these reasons, large donor-funded schemes are being implemented. From our observation, people are found more concerned about distributive justice, i.e., how access to water can be ensured, over procedural and recognitive justice issues. Though some of the problems of representation and participation in the users committee were raised, the major concern was equitable access and distribution of water.

In comparison, the issue of procedural justice was more of a concern in Dhulikhel in addition to the dominance of distributive justice concerns. The increasing demand regarding procedural justice is due to growing awareness and participation of women towards inclusion and progressive national policies, for example, a mandatory quota for women in civil service and legislative bodies. The Dhulikhel drinking water users committee is a traditional and established community-based water management system, operating since the 1990s with a progressive representation of women in the committee. In Diktel, the water users committees are reformed lately and evolving with new project development. As a town nearby Kathmandu, Dhulikhel women are building on their agency to voice their concerns, identifying the unjust practices and decisions of patriarchal society.

In contrast, recognitive justice concerns were increasingly raised in Diktel compared to Dhulikhel because there is growing awareness of recognitive rights, particularly among the indigenous ethnic groups. Issues of distributive justice dominate people's perceptions and claims of justice. For them, getting access to water forms the priority and fundamental need to be ensured, after which come the other two forms of justice. However, recognitive and procedural justice are also crucial as these justice dimensions help people address their distributive justice concerns.

Discussion and Conclusion

This chapter analyzed various environmental and climate justice issues by focusing on three key justice dimensions. We situated these justice concerns in the context of social differentiation in Nepal according to class, caste, ethnicity, gender, and spatiality. The chapter highlighted the limited climate justice debate in the country. Using a climate justice framework, we analyzed two urban water governance case studies, Dhulikhel and Diktel, amid increasing climate change problems. These case studies show inequities regarding access to water services, participation of local communities in water-related decision-making processes, and problems of recognition. We also showed how, for most water users, issues of distributive justice, for example, access to water services, remain the dominant concern over procedural and recognition concerns.

Our case studies highlighted how the large-scale water supply schemes designed to address increasing urban water insecurity in the Himalayan towns because of the climate change crisis actually worsen the spatial inequity in water distribution. Above all, our studies show that these projects bear the risk of unequal water distribution and access (Chowdhury & Rasul, 2011; Kovács et al., 2019). We illustrated that large-scale adaptation interventions could marginalize the poorer sections of society when traditional structures are disregarded and dismantled. Likewise, impoverished communities in higher elevations cannot meet their required water demand due to their unaffordability to invest in technology. Distributive justice weaves together inequities of the class, caste, ethnicity, gender, and spatiality.

We also demonstrated that poor sections of society face exclusion from the key decision-making structures and processes related to water governance. The development agencies are often found to pose control over water resources and water-related decision-making (Leder et al., 2019). The domination of powerful actors in water governance mechanisms leaves women, the poor, and marginalized groups excluded or with limited access to decision-making. Thus, more attention to equitable distribution and participatory approaches is necessary.

Furthermore, technical fields such as water projects construction and management are generally dominated by men. Women's voices are rarely heard in water governance (Bhattarai et al., 2021). We also find that claims by upstream communities about water raise concerns of recognitive justice. There should be better recognition of the rights of the local communities and ethnic groups residing in the water source area, as well as a clear institutional mechanism addressing customary upstream-downstream claims.

Among the three dimensions of justice, we highlighted that distributive justice is the most important problem in Dhulikhel and Diktel. Besides distributive justice concerns, procedural justice concerns were also raised in Dhulikhel, while recognitive justice was featured in Diktel. These findings align with Sikor et al. (2019), who studied a hydropower project's justice dimensions in Western Nepal. People there were more concerned with distributive justice regarding compensation, infrastructure development, remedies against dust and other losses, and benefits (Sikor et al., 2019). Other studies have indicated that recognitive justice is the least well-understood aspect of climate justice despite its high relevance in conservation owing to local knowledge and cultures (Martin et al., 2016).

Our application of a climate justice framework highlights the usefulness of the framework to identify inequities in climate change adaptation and disaster risk management. A similar framework can be applied to illustrate the differential impacts of climate-induced disasters on different social groups and how effectively and equitably disaster risk responses and management decisions are handled. Disaster-related stress can be overwhelming, and during such distress pre-existing vulnerabilities and exclusion (including caste inequality) often become more widespread.

Acknowledgments

The authors would like to acknowledge the projects 'Climate Adaptive Water Management Plans for Cities in South Asia' Grant Number: 108212-001, funded by IDRC for the PhD programme and Governing climate-resilient futures: Gender, justice and conflict resolution in resource management (JustClime) funded by the Swedish Research Council.

Notes

1 Dhulikhel is one of the three major urban centres in the Kavre valley.
2 It consists of acknowledging actors' distinct identities and histories and eliminating the cultural domination of some groups over others.
3 It is expected to supply water in the core town of Diktel, covering only 10.4% of its population, i.e., water supply to 1,049 out of 10,050 households.
4 22 community taps of town drinking water and 32 taps of village drinking water projects.
5 For instance, she shared about the last-minute change of a guest invitee in one of the events by the chair in consultation with other male members but without her knowledge.
6 Such forums were practised for discussing water-related issues and opting for solutions (1 in Diktel and 7 in Dhulikhel were held).

References

Adger, W. N., Paavola, J., Huq, S., & Mace, M. (2006). Toward Justice in Adaptation to Climate Change. *Fairness in Adaptation to Climate Change*, 1–19.

Berrang-Ford, L., Siders, A., Lesnikowski, A., Fischer, A. P., Callaghan, M. W., Haddaway, N. R., Mach, K. J., Araos, M., Shah, M. A. R., & Wannewitz, M. (2021). A Systematic

Global Stockpile of Evidence on Human Adaptation to Climate Change. *Nature Climate Change*, *11*(11), 989–1000.
Bhattarai, B., Beilin, R., & Ford, R. (2015). Gender, Agrobiodiversity, and Climate Change: A Study of Adaptation Practices in the Nepal Himalayas. *World Development*, *70*, 122–132.
Bhattarai, B., Upadhyaya, R., Neupane, K. R., Devkota, K., Maskey, G., Shrestha, S., Mainali, B., & Ojha, H. (2021). Gender Inequality in Urban Water Governance: Continuity and Change in Two Towns of Nepal. *World Water Policy*, *7*(1), 30–51.
Chakraborty, R., & Sherpa, P. Y. (2021). From Climate Adaptation to Climate Justice: Critical Reflections on the IPCC and Himalayan Climate Knowledges. *Climatic Change*, *167*(3–4), 49.
Chowdhury, A., & Rasul, G. (2011). Equity and Social Justice in Water Resource Governance: The Case of Bangladesh. *South Asian Water Studies*, *2*(2), 44–58.
Duncan, J. M., Biggs, E. M., Dash, J. and Atkinson, P. M. (2013). Spatio-temporal Trends in Precipitation and their Implications for Water Resources Management in Climate-Sensitive Nepal. *Applied Geography*, *43*, 138–146. https ://doi.org/10.1016/j.apgeo g.2013.06.011.
Eriksen, S., Schipper, E. L. F., Scoville-Simonds, M., Vincent, K., Adam, H. N., Brooks, N., Harding, B., Lenaerts, L., Liverman, D., & Mills-Novoa, M. (2021). Adaptation Interventions and their EFFECT on Vulnerability in Developing Countries: Help, Hindrance or Irrelevance? *World Development*, *141*, 105383.
Gautam, M. R., Timilsina, G. R., & Acharya, K. (2013). Climate Change in the Himalayas: Current State of Knowledge. *World Bank Policy Research Working Paper*, (6516).
Gentle, P., Thwaites, R., Race, D., & Alexander, K. (2014). Differential Impacts of Climate Change on Communities in the Middle Hills Region of Nepal. *Natural Hazards*, *74*, 815–836.
Gerlitz, Y. J., Brook, N., Macchi, M., Pandey, R., & Jha, S. (2015). The Multidimensional Livelihood Vulnerability Index: An Instrumental to Measure Livelihood Vulnerability to Change in the Hindu Kush Himalayas. In A. Markandya and D. Rubbelke (Eds.) *Climate and Development*. New Jersey: World Scientific.
Gulliver, R., Vachette, A., & Boddington, S. (2023 07 August). How Australian Environmental Non-Governmental Organisations Frame and Enact Climate Justice. NPJ Climate Action 2: 18.
Henrique, K. P., & Tschakert, P. (2021). Pathways to Urban Transformation: From Dispossession to Climate Justice. *Progress in Human Geography*, *45*(5), 1169–1191.
Homberg, M. V. D., & McQuistan, C. (2019). Technology for Climate Justice: A Reporting Framework for Loss and Damage as Part of Key Global Agreements. In S. Surminski, J. Linnerooth-Bayer, & L. Bouwer (Eds.) *Loss and Damage from Climate Change: Concepts, Methods and Policy Options* (pp. 513–545). Cham: Springer.
Honneth, A. (2001). Recognition or Redistribution? *Theory, Culture & Society*, *18*(2–3), 43–55.
Islam, N., & Winkel, J. (2017). Climate Change and Social Inequality. U.N. Department of Economic and Social Affairs Working Paper No. 15. New York: United Nations.
Kovács, E. K., Ojha, H., Neupane, K. R., Niven, T., Agarwal, C., Chauhan, D., Dahal, N., Devkota, K., Guleria, V., Joshi, T., Michael, N. K., Pandey, A., Singh, N., Singh, V., Thadani, R., & Vira, B. (2019). A Political Ecology of Water and Small-town Urbanisation across the Lower Himalayas. *Geoforum*, *107*, 88–98. https://doi.org/10.1016/j.geoforum.2019.10.008
Leder, S., Shrestha, G., & Das, D. (2019). Transformative Engagements with Gender Relations in Agriculture and Water Governance. *New Angle: Nepal Journal of Social Science and Public Policy*, 5(1), 128–158.

Macchi, M., Gurung, A. M., & Hoermann, B. (2015). Community Perceptions and Responses to Climate Variability and Change in the Himalayas. *Climate and Development*, *7*(5), 414–425.

Madhanagopal, D., Dahal, S., & Nikku, B. R. (2022) Responding to Environmental Disasters in India and Nepal: Insights from Green Social Work. In D. Madhanagopal & B. Nikku, *Social Work and Climate Justice*, London: Routledge. 172–190.

Manandhar, U., & Bhatta, G. D. (2013). Biogas for Climate Justice: A Story of Change in Nepal. In Irish Aid Programme. A new dialogue: putting people at the heart of global development. Papers of the Hunger, Nutrition and Climate Justice Conference, Dublin, Ireland, 15-16 April 2013. Dublin, Ireland: Irish Aid Programme. pp. 39–42.

Marino, E., & Ribot, J. (2012 May). Special Issue Introduction: Adding Insult to Injury: Climate Change and the Inequities of Climate Intervention *Global Environmental Change*, 22(2), 323–328.

Martin, A., Coolsaet, B., Corbera, E., Dawson, N. M., Fraser, J. A., Lehmann, I., & Rodriguez, I. (2016). Justice and Conservation: The Need to Incorporate Recognition. *Biological Conservation*, 197, 254–261.

Maskey, G., Pandey, C., Bajracharya, R. M., & Moncada, S. (2021). Inequity in Water Distribution and Quality: A Study of Mid-Hill Town of Nepal. *World Water Policy*, *7*(2), 233–252.

McSweeney, C., New, M., & Lizcano, G. (2011). UNDP Climate Change Country Profiles: Nepal. Available at https://climateknowledgeportal.worldbank.org/sites/default/files/2018-10/wb_gfdrr_climate_change_country_profile_for_NPL.pdf (accessed on September 24, 2018).

Ministry of Environment (MoE) (2010). *Climate Change Vulnerability Mapping for Nepal*. Kathmandu: MoE.

MoHA (2013). *Nepal Disaster Report 2013: Focus on Participation and Inclusion*. Kathmandu: Ministry of Home Affairs (MoHA), the Government of Nepal.

Newell, P., Srivastava, S., Naess, L. O., Torres Contreras, G. A., & Price, R. (2021). Toward Transformative Climate Justice: An Emerging Research Agenda. *Wiley Interdisciplinary Reviews: Climate Change*, *12*(6), e733.

O'Brien, K. (2016). Climate Change Adaptation and Social Transformation. In D. Richardson, *International Encyclopedia of Geography: People, the Earth, Environment and Technology: People, the Earth, Environment and Technology* (pp. 1–8). Chichester: John Wiley and Sons.

Pelling, M. (2010). *Adaptation to Climate Change: From Resilience to Transformation*. London: Routledge.

Pokharel, K., Joshi, T., Adhikari, J., & Neupane, K. R. (2018). Water Security in Urban Water Governance: A Case Study of Dhulikhel Municipality, Kavrepalanchok, Nepal.

Porter, L., Rickards, L., Verlie, B., Bosomworth, K., Moloney, S., Lay, B., Latham, B., Anguelovski, I., & Pellow, D. (2020). Climate Justice in a Climate Changed World. *Planning Theory & Practice*, *21*(2), 293–321.

Poudel, S., Funakawa, S., Shinjo, H., & Mishra, B. (2019). Understanding Households' Livelihood Vulnerability to Climate Change in the Lamjung District of Nepal. *Environment, Development and Sustainability,* 22(10).

Poudel, S., & Shaw, R. (2016). The Relationships between Climate Variability and Crop Yield in a Mountainous Environment: A Case Study in Lamjung District, Nepal. *Climate*, *4*, 13.

Pradhan, P., Seydewitz, T., Zhou, B., Lüdeke, M. K., & Kropp, J. P. (2022). Climate Extremes Are Becoming More Frequent, Co-Occurring, and Persistent in Europe. *Anthropocene Science*, *1*(2), 264–277.

Pradhan, P., Subedi, D. R., Khatiwada, D., Joshi, K. K., Kafle, S., Chhetri, R. P., Dhakal, S., Gautam, A. P., Khatiwada, P. P., & Mainaly, J. (2021). The COVID-19 Pandemic Not Only Poses Challenges, but Also Opens Opportunities for Sustainable Transformation. *Earth's Future*, *9*(7), e2021EF001996.

Radonic, L., & Zuniga-Teran, A. (2023). When Governing Urban Waters Differently: Five Tenets for Socio-Environmental Justice in Urban Climate Adaptation Interventions. *Sustainability*, *15*(2), 1598.

Routledge, P., Cumbers, A., & Derickson, K. D. (2018). States of Just Transition: Realising Climate Justice through and against the State. *Geoforum*, *88*, 78–86.

Satyal, P. (2011). Shifting Conceptions of Social (in) Justice in Nepal. *New Angle: Nepal Journal of Social Science and Public Policy*, *1*(1), 49–64.

Satyal, P., Byskov, M. F., & Hyams, K. (2021). Addressing Multi-Dimensional Injustice in Indigenous Adaptation: The Case of Uganda's Batwa Community. *Climate and Development*, *13*(6), 529–542.

Satyal, P., Corbera, E., Dawson, N., Dhungana, H., & Maskey, G. (2020). Justice-Related Impacts and Social Differentiation Dynamics in Nepal's REDD+ Projects. *Forest Policy and Economics*, *117*, 102203.

Schlosberg, D. (2007). *Defining Environmental Justice: Theories, Movements, and Nature*. Oxford: Oxford University Press.

Shrestha, A., Maskey, G., & Basnet, S. (2022). Preserving the Mid-Hill Springs. Blog published in The Rising Nepal. Available at: https://www.risingnepaldaily.com/news/14206

Shrestha, B. A., & Aryal, R. (2011). Climate Change in Nepal and Its Impact on Himalayan Glaciers. *Regional Environmental Change*, *11*, 65–77.

Shrestha, K. K., Bhattarai, B., Ojha, H. R., & Bajracharya, A. (2019). Disaster Justice in Nepal's Earthquake Recovery. *International Journal of Disaster Risk Reduction*, *33*, 207–216.

Sikor, T., Fisher, J., Few, R., Martin, A., & Zeitoun, M. (2013). The Justices and Injustices of Ecosystem Services. In T. Sikor (Ed.) *The Justices and Injustices of Ecosystem Services* (pp. 201–214). Milton Park: Routledge.

Sikor, T., Satyal, P., Dhungana, H., & Maskey, G. (2019). Brokering Justice: Global Indigenous Rights and Struggles over Hydropower in Nepal. *Canadian Journal of Development Studies/Revue canadienne d'études du développement*, *40*(3), 311–329.

Thomas, K., Hardy, R. D., Lazrus, H., Mendez, M., Orlove, B., Rivera-Collazo, I., Roberts, J. T., Rockman, M., Warner, B. P., & Winthrop, R. (2019). Explaining Differential Vulnerability to Climate Change: A Social Science Review. *Wiley Interdisciplinary Reviews: Climate Change*, *10*(2), e565.

Walker, G. (2012). *Environmental Justice: Concepts, Evidence and Politics*. Hoboken: Taylor and Francis.

Ziervogel, G., Cowen, A., & Ziniades, J. (2016). Moving from Adaptive to Transformative Capacity: Building Foundations for Inclusive, Thriving, and Regenerative Urban Settlements. *Sustainability*, *8*(9), 955.

Part 7

Health Equity

22 The Stress of Poverty in Tackling Tuberculosis in Nepal

Marissa Taylor

When his daughter developed a light cough sometime in the spring of 2021, Rameshwor Maharjan was not particularly worried. His 21-year-old daughter Rezina had tested negative for Covid-19 on her PCR test and she was a perfectly healthy girl, so he supposed the cough was just the flu and that it would go away in a few days.

But it persisted for weeks. Although anxiousness had started to creep in, Rameshwor thought it'd be best to stay away from hospitals, as the country was then under a nationwide lockdown and hospitals were filled with Covid-19 patients. Hospitalizations had skyrocketed at the time with the country reeling under an acute shortage of oxygen as well (The Kathmandu Post, 2021).

"Not taking my daughter immediately to see a doctor was my biggest mistake," said 51-year-old Rameshwor, a resident of Lalitpur Metropolitan City in Bagmati province.

Less than six months later, Rezina died. The light cough wasn't the flu, nor was it Covid-19 – it turned out to be pulmonary tuberculosis. When Rameshwor finally took her for a check-up in June, almost two months after the first symptoms appeared, doctors said they had come in late. Bacteria had almost completely infected her left lung.

She was immediately put on a six-month antibiotics course, which is the standard procedure to treat patients with pulmonary tuberculosis. Rezina had just one week left to complete her course, says Rameshwor, but the medicines were not doing their job.

"She was just not getting better. Her immunity had collapsed completely. She had difficulty just moving her body; it left her exhausted. She did not even have the energy to get out of bed," he said.

Then one day, she started vomiting blood. "Soon after, blood was coming out of her nose too. She died around 15 minutes after we'd reached the hospital," said Rameshwor.

Every year, millions like Rezina succumb to tuberculosis, or TB. In the year 2020 alone, the disease killed 1.3 million around the world (World Health Organization, 2021a, 2021b) with a quarter of the world's total population carrying the TB bacteria (Houben et al., 2016).

DOI: 10.4324/9781003371175-29
This chapter has been made available under a CC-BY-NC-ND 4.0 license.

TB is a major health problem globally, ranking as the ninth leading cause of mortality. It causes more deaths than any other infectious disease in the world today, surpassing HIV/acquired immune deficiency syndrome (Christopher, 2019). In Nepal too, the disease claims thousands of lives every year: an estimated 17,000 people die due to the disease every year and over 117,000 Nepalis are currently living with the disease (National Tuberculosis Control Centre, 2019a, 2019b).

"Tuberculosis is the leading cause of death from a single infectious agent. It is an epidemic in Nepal," said Dr Naveen Prakash Shah, a senior consultant with the National Tuberculosis Control Center, which runs the country's Tuberculosis Control Programme in Nepal.

However, despite the disease's high prevalence and mortality rate, health systems around the world have not been able to build systems robust enough to control the disease's spread. TB is a particular problem in low-income countries like Nepal, where a majority of people do not have access to basic health care, and in extremely poor environments with high population density – two major components that foster the bacteria that cause tuberculosis to survive, thrive and kill.

Tuberculosis is one of the oldest diseases in the history of mankind, with evidence of tubercular decay found in some Egyptian mummies from 3000 to 2400 BC (Amer & Mohammed, 2013). And throughout history, the disease has been a major health crisis. Today, however, in many rich, developed countries, the disease has been contained. But in Nepal and other developing nations, it remains a serious threat.

"Tuberculosis is such a prevalent disease in Nepal because it is 'a poor man's disease'," says Dr Sushil Baral, a health and development expert who specializes in strengthening health systems by building robust health policies at the national and international levels. He has played a significant role in creating Nepal's National Tuberculosis Strategic Plan. Dr. Baral described this problem as one of social equity.

> I say that TB is a disease of the poor because areas with the highest number of TB deaths are in the developing world, among populations belonging to lower income backgrounds who do not have access to proper nutrition, compromising their immunity and making them susceptible to the bacteria,

Unprecedented growth of the urban population, especially the urban poor, has increased the concentration of at-risk populations in Nepal, specifically in the Kathmandu Valley and in urban areas of the Tarai lowlands. Among the registered TB cases, 24% were reported in Province 3, of which 10% were from Kathmandu district (National Tuberculosis Center, 2019a).

Nepal is not the only country where poor environmental conditions created a high TB burden. In neighboring India too, the case is the same. The TB prevalence is 60% higher among people living below the poverty line compared with those above the poverty line. Among marginalized people, TB was 1.5 times more prevalent. In fact, TB strikes the poor at disproportionately high rates across South Asia (Muniyandi et al., 2008).

Poverty leads to poor health, which in turn aggravates poverty. In fact, poverty directly accounts for almost one-third of the global burden of the disease (Benatar & Upshur, 2010). Ninety-five percent of TB cases and 98% of TB deaths are in developing countries.

Krishna Gopal Prajapati, 62, has worked as a farmer his whole life. Every day for the past five decades, he has gone out to his fields in Bhaktapur, a district adjacent to the capital of Kathmandu, and worked all day. Every morning before he headed out to the fields, he would drink a small glass of alcohol. During the day, he would take another drink during a break. A cigarette and a swig of some local alcohol always followed his lunch. At the end of the day, back from after toiling in the fields all day, he would take a couple of glasses more.

"You can't work in the fields if you don't drink," said Prajapati, "And I needed to smoke while I drank."

Every day for almost five decades, Prajapati says he has had at least one drink and smoked a packet of cigarettes. Two months ago, he found out he has pulmonary tuberculosis, a disease he had no idea existed before he got diagnosed.

Smoking tobacco is perhaps the number one reason behind TB. It harms both the smoker and those who inhale second-hand smoke, as it reduces immune system strength, but people like Prajapati, who are uneducated and live in poverty, are unaware of just how dangerous smoking can be.

He is now in the second month of his six-month antibiotics course, and his cough is getting better. But he lives with his daughter and wife in a small two-room flat. Since he got diagnosed, he has isolated himself in one of the rooms but it's difficult to isolate oneself in such a small space.

"I am worried I will give the disease to my daughter," he says.

People with active TB can infect 5–15 other people through close contact over the course of a year (World Health Organization, 2018). Overcrowded living conditions often increase the chance of TB transmission and worsen outcomes. In fact, urbanization combined with poverty creates the perfect conditions to increase TB transmission. This is because urbanization and poverty lead to greater population density, more crowded, poorly ventilated living conditions, and increased mobility of people – all of which adds to the disease's spread.

Another environmental condition that causes TB is sustained exposure to air pollution.

In recent years, air pollution in Nepal has become a serious environmental and public health concern, with pollution levels 4.9 times higher than recommended by the World Health Organization. This deteriorating air quality has put hundreds of thousands of Nepalis at risk of several health problems, especially in clustered urban areas like Kathmandu.

Globally, poor air quality is estimated to cause some 7 million deaths each year, as poor air increases the risk of contracting a wide range of cardiovascular and respiratory diseases. In Nepal too the situation is dire: in 2019, the country had the world's highest age-standardized death rates for chronic lung disease caused by air pollution – 182.5 per 100,000 population with 3318.4 years lost due to ill health or disability (Safiri et al., 2022).

This exposure to poor air and the impact of air pollution are not equally distributed. Poorer and marginalized communities are often more exposed to areas with higher levels of pollution because pollution sources, such as industrial facilities or sewage corridors, are disproportionately located in low-income neighborhoods. In addition, the lack of affordable quality healthcare services further increases air pollution-related mortality (Lelieveld et al., 2020).

Epidemiological research also shows that TB is more prevalent in people exposed to indoor air pollution in homes where, for example, firewood or charcoal is used to cook. In Nepal, close to two-third of the population still use solid fuels such as firewood (64%) as a primary source of energy, with around 8,700 people dying prematurely each year due to illnesses related to indoor air pollution from solid biomass fuel burning. As women play a significant role in household cooking, they are disproportionately affected by indoor air pollution and inadvertently by TB. Dr. Janak Koirala, an infectious disease expert observed,

> Long-term exposure to dust or smoke, especially from wood fire, particularly for women, can cause damage to lungs and develop diseases like chronic bronchitis and chronic obstructive pulmonary disease. This weakens the body's natural defense system in the lungs and can make people susceptible to TB,

And things are only becoming grimmer: the bacteria that causes TB is developing resistance to the antibiotics that are used to kill it.

Tuberculosis is a bacterial disease that usually harms the lungs, but can also attack other parts of the body, like the lymph nodes, kidneys, bones, spine, and even the brain. It is airborne, meaning if you inhale the air around someone hosting various strains of mycobacteria, the bacteria that causes the disease, you will likely be infected.

But not everyone infected with the TB bacteria becomes sick. The bacteria can live in the human body without making you ill. With these "latent" TB infections, the body is able to fight the bacteria to stop them from manifesting the disease.

However, if the bacteria enters your body and your body's immune system cannot fight it, the bacteria start attacking your body. In the case of pulmonary TB, it attacks the lungs and digs holes in the lung tissue, creating respiratory complications, lung damage, and eventually, death. Usually, the first symptoms of cough, fever, loss of appetite, weight loss, and extreme tiredness surface only a couple of weeks after the bacteria enter your body – after the bacteria have done enough damage, making the disease tricky to detect and thus dangerous.

But as alarming as TB sounds, it can be easy to treat, thanks to antibiotics. With a course of four standard, or first-line, anti-TB drugs most TB patients can be cured.

But these days the bacteria present in the lungs of a TB patient have developed resistance to these antibiotics. As a result, TB is becoming much harder to treat, increasing the risk of disease spread.

Antibiotics were developed in the 1940s to prevent and treat bacterial infections in humans, animals, and plants. But when these drugs are overused, as they are in Nepal and around the world making way for a silent epidemic, the bacteria they were designed to kill start building resistance toward them. This phenomenon is being seen not just in bacteria but in other microbes like fungi, viruses, and parasites as well. This phenomenon is known as Antimicrobial Resistance, or AMR.

"Antimicrobial resistance is a global health threat, and it will soon be the leading cause for mortality, even among treatable diseases," said Adarsh Man Sherchan, a conservation geneticist and molecular consultant who is currently studying Antimicrobial Resistance loads in the Kathmandu Valley's Bagmati River.

"There are several reasons why AMR continues to emerge and spread. Chief among them is haphazard consumption of antibiotics with a history of self-medication and premature interruption of treatment," said Sherchan.

This resistance has given rise to new strains of TB, such as Drug-Resistant tuberculosis (DR-TB), Multidrug-Resistant tuberculosis (MDR-TB), and Extensively Drug-Resistant tuberculosis (XDR-TB). All are resistant to some of the most effective anti-TB drugs.

DR-TB is a major contributor to antimicrobial resistance worldwide and continues to be a public health threat. Annually, about half a million people fall ill with drug-resistant TB. It is also making TB harder and longer to treat, often with poorer outcomes for patients.

With a majority of the population living in poverty and poor environmental conditions, it is not surprising that Nepal is among the 30 countries where the burden of DR-TB is high. An estimated 2,200 cases of Drug-Resistant TB and 450 MDR-TB cases are identified annually in Nepal. These are just estimates and the actual number is possibly much higher (World Health Organization, 2021a).

Detection of DR-TB requires various tests, such as rapid molecular tests, culture method tests, and sequencing technologies. But such services are not available throughout Nepal, causing severe gaps in detection. According to the National Tuberculosis Control Center, there are only 22 DR-TB treatment centers across the country.

> Say I live in Bhojpur. To do a culture or drug sensitivity test, I will have to go to Dharan to the BP Koirala Institute of Health Sciences because health posts and even district hospitals do not have such sophisticated technologies,

said Dr Baral.

> And while treatment is free at government hospitals, in the process, all indirect costs have to be borne by the patient and the family. Many are not able to afford it and thus do not seek the treatment they need. In this way, many cases never make it to our records. This is extremely dangerous, as this means an untreated patient with active DR-TB bacteria could be infecting others,

said Baral.

This lack of availability and access to early screening of TB cases is the primary reason why DR-TB cases have remained mostly entrenched in Nepal.

"If the government does not take immediate action, then this will blow up, because treatment is costly and if patients are not treated in a timely fashion and tracked then it will continue to spread," said Dr Baral.

So, what are governments around the world doing about TB and how does Nepal compare? The World Health Organization's End TB Strategy (2016–2035) aims to end the global tuberculosis epidemic by 2035. The strategy includes targets to reduce absolute mortality by 95% and incidence by 90% between 2015 and 2035. The strategy was adopted with full support by the World Health Assembly in Geneva in May 2014 and urges governments to offer financing and high-level commitment to facilitate the strategy's implementation.

While the strategy's long-term vision is to eliminate tuberculosis globally, defined as less than one new tuberculosis case per million people per year, executing the pledge to "end the global tuberculosis epidemic" would mean reducing the global incidence rate from greater than 1,000 new cases per million people in 2015 to less than 100 new cases per million people by 2035.

In Nepal, efforts to reduce TB cases were introduced long ago. The Tuberculosis Control Programme was launched by the Government of Nepal almost six decades ago. To control the spread of TB and the various strains of drug-resistant TB, the government also developed a National Strategic Plan to End Tuberculosis in Nepal and a Tuberculosis Free Nepal Declaration Initiative Implementation Guideline, which aims to end TB by 2035 and put in place better testing facilities.

Currently, the state provides free treatment through 135 public hospitals, 2,168 non-government organizations, 196 Primary Health Care Centers, 3,806 health posts, and 51,420 Female Community Health Volunteers. A total of 4,244 DOTS treatment centers are providing TB treatment service across the country. Why then is the national TB burden still so high?

A major problem is a gap in case detection.

> There is a huge gap in detection on the part of the government, which shows that access to healthcare is compromised. According to the government's own estimates, there are 69,000 new TB cases annually but only 27,000 people have actually been diagnosed and have sought treatment,

said Dr Baral. "Where are the rest of the people? Why has the government been unable to track them?"

Early identification is crucial to break the chain of transmission. But owing to a weak national surveillance program – a result of inadequate manpower, infrastructure, and budget – that has been an impossible task.

But it is only through early-stage intervention that the TB problem can be solved, experts say. "Our treatment success rate is 90% and completely free to the public, which is something to be proud of. But our focus now has to shift toward reducing the number of cases, toward nipping the problem at the bud," said Ram Sharan Gopali, the director of the Japan-Nepal Tuberculosis Research Association, which is one of the organizations partnering with the government of Nepal as a part of the national TB program. The organization oversees the government's efforts in Gandaki and Bagmati provinces.

Cases like Rezina's are all too common in Nepal, where patients reach hospitals too late. "In between trying to figure out what the disease is and getting treated for pneumonia, people come in to seek treatment too late, and in the process infect others as well," Gopali said.

Awareness is another major challenge.

"There is a general lack of awareness still among people regarding tuberculosis and the government's efforts towards education have not been enough," he said.

The level of TB awareness is associated with various demographic and socio-economic factors such as education level, socioeconomic status, and area of residence. Low public awareness of TB has been identified as correlated with several socioeconomic factors including family income, education level, and gender. In addition, enhanced awareness of TB could improve access to appropriate treatment and better outcomes, especially in socioeconomically vulnerable groups – who have limited access to information about TB due to weak community engagement – and certain areas of residence (Baral et al., 2007).

In lower-middle income countries like Nepal, specific locations and places at greater distances from TB treatment facilities are susceptible to poor community TB awareness and weak adherence to TB treatment. Examining the demographic and socioeconomic factors associated with TB awareness is important for TB control.

> TB is an old disease, it somehow doesn't get the attention it deserves. It is not as glamorous or dangerous as others. TB is a major, major global health crisis. And it is so because it is not about fighting a single disease. To fight tuberculosis, we have to fight illiteracy, poverty, unequal access to health care and social stigma,

said Dr Baral. "To eliminate the disease, we need a multi-sectoral approach to fight it. And that is a huge challenge."

References

Amer, S., & Mohammed, A. (2013). The Next Pandemic - Tuberculosis: The Oldest Disease of Mankind Rising One More Time. *British Journal of Medical Practitioners*. https://www.bjmp.org/content/next-pandemic-tuberculosis-oldest-disease-mankind-rising-one-more-time#:~:text=Tuberculosis%20is%20one%20of%20the,the%20Greek%20term%20for%20tuberculosis

Baral, S. C., Karki, D. K., & Newell, J. N. (2007). Causes of Stigma and Discrimination Associated with Tuberculosis in Nepal: A Qualitative Study. *BMC Public Health*, 7, 1–10. https://link.springer.com/content/pdf/10.1186/1471-2458-7-211.pdf?pdf=button

Benatar, S. R., & Upshur, R. (2010). Tuberculosis and Poverty: What Could (and should) Be Done?. *International Journal of Tuberculosis and Lung Disease*. https://pubmed.ncbi.nlm.nih.gov/20843410/

Christopher, A. J. (2019). Realizing the World Health Organization's End TB Strategy (2016–2035): How Can Social Approaches to Tuberculosis Elimination Contribute to Progress in Asia and the Pacific?. *MDPI.* https://www.mdpi.com/2414-6366/4/1/28#B2-tropicalmed-04-00028

Houben, R. M. G. J., & Dodd, J. P. (2016). The Global Burden of Latent Tuberculosis Infection: A Re-estimation Using Mathematical Modelling, National Centre for Biotechnology Information. https://www.ncbi.nlm.nih.gov/pmc/articles/PMC5079585/

Lelieveld, J., Pozzer A., Pöschl, U., Fnais M., Haines, A., & Münzel, T. (2018). *Loss of Life Expectancy from Air Pollution Compared to Other Risk Factors: A Worldwide Perspective.* Cardiovascular Research. Oxford Academic (oup.com)

Muniyandi, M., & Ramachandran, R. (2008). Socioeconomic Inequalities of Tuberculosis in India. Expert Opinion on Pharmacotherapy. https://www.tandfonline.com/doi/full/10.1517/14656566.9.10.1623

National Tuberculosis Control Centre (2019a). National TB Prevalence Survey. https://nepalntp.gov.np/prevalence-survey/

National Tuberculosis Control Centre (2019b). National Tuberculosis Program Nepal. https://nepalntp.gov.np/prevalence-survey/

Poudel, A. (April 20, 2021). Most Hospitals Do Not Have Ready Supply of Oxygen. The Kathmandu Post. https://kathmandupost.com/health/2021/04/20/new-cases-are-exploding-but-most-hospitals-do-not-have-ready-supply-of-oxygen

Safiri, S., Carson-Chahhoud, K., Noori, M., Nejadghaderi, S. A., & Sullman, M. J. M. (2022). Burden of Chronic Obstructive Pulmonary Disease and Its Attributable Risk Factors in 204 Countries and Territories, 1990–2019: Results from the Global Burden of Disease Study 2019. The BMJ. https://www.bmj.com/content/378/bmj-2021-069679

World Health Organization (2018). World Health Organization Website. https://www.who.int/news-room/questions-and-answers/item/tuberculosis#:~:text=People%20ill%20with%20TB%20can,the%20course%20of%20a%20year

World Health Organization (2021a). Global Tuberculosis Report 2021. https://www.who.int/publications/i/item/9789240037021

World Health Organization (April, 2021b). World Health Organization website. WHO Releases New Global Lists of High-Burden Countries for TB, HIV-Associated TB and Drug-Resistant TB.

23 Impacts of Lead Contamination on Children's Health in Nepal

Meghnath Dhimal, Mandira Lamichhane Dhimal, and Madhusudan Subedi

Introduction

Lead, widely used in many industries and consumer products, is a highly poisonous element that affects almost every organ in the body. Exposure to lead mainly occurs through inhalation or ingestion of food, paint, water, soil, or dust. Lead accumulates in the body in blood, bone, and soft tissue. There is no safe level of lead in the body.

Infants, young children, adolescents, and pregnant women are more susceptible to the adverse health effects of lead. Children ingest more food and water and breathe more air in relation to their body size than adults. Furthermore, due to their age, children do not have understanding about avoiding contaminated food and drinks: they often put dirt and other things into their mouths. Adverse environmental conditions and pollution are major contributors to childhood deaths, illnesses and disability, particularly in developing countries like Nepal.

Although it took decades to phase out lead in petrol, by the end of 2005, nearly all Low and Middle Income Nations (LMICs) had done so. Even so, a large and growing body of research indicates that lead exposure in some LMICs has increased or is still significant.

Average blood lead levels (BLL) in the general population in LMICs are significantly higher than those in high-income countries (Ericson et al., 2021) and there are also inequalities within national populations among LMICs. A UNICEF report stated that up to 800 million children worldwide, or around one in every three children, had BLL at or above 5 micrograms per deciliter (g/dL) (Olufemi et al., 2022).

Children who are exposed to lead face long-lasting health and developmental effects including the brain and nervous system damage. Lead exposure can also cause slower growth and development such as difficulty in learning, hearing, as well as behavior and speech issues. Lower IQ, a loss of focus and concentration, and poor academic achievement are all possible consequences of lead's neurological effects. The likelihood of miscarriage, early birth, and other adverse effects on the development of the fetus is also heightened (Dhimal et al., 2017; Ericson et al., 2021; Kosnett et al., 2007; Olufemi et al., 2022; Zamani et al., 2023).

Prevention is the only effective way to ward off the damage caused by lead poisoning. Keeping lead-based products out of reach of children, maintaining proper

DOI: 10.4324/9781003371175-30
This chapter has been made available under a CC-BY-NC-ND 4.0 license.

hygiene, regular hand washing, and environmental sanitation are the major ways to reduce lead exposure.

Lead contamination is rampant in Nepal. In this chapter, we describe the sources of lead contamination, its impact on children, and resulting environmental health injustice. In addition, we recommend measures to prevent health risks of lead exposure in the context of Nepal. We argue that the effects of lead contamination on children's health is a major environmental justice crisis for Nepal and one in need of immediate and large-scale action by government, non-government agencies, and corporate sectors.

We find that the increasing pattern of urbanization, industrialization, generation of municipal and electronic waste, use of enamel paints in buildings, and other factors has increased exposure to lead, especially in Nepal's urban centers. Children of poor and socially marginalized groups face a disproportionately high rate of lead exposure. Lead exposure during childhood causes cognitive deficiencies as well as a decline in social mobility. Household exposure, environmental exposure, and other various factors such as socioeconomic status, malnutrition, poverty, poor education, and certain cultural practices are strongly associated with high blood lead level, especially in younger children. Among adults, limited labor rights, insufficient workforce protections, and lack of work insecurity in the industry have caused increased lead exposure. In order to address such problems, the governments in Nepal must implement effective regulations and policies to help maintain workplace safety and industrial hygiene as well as clean up environments where children live and play.

Study Area

We focus particularly on Nepal's urban areas as they have the greatest threats to lead exposures for children. In 2021, the urban population of Nepal reached 66.8% of the country's total population of 29 million with Kathmandu Metropolitan City recording the highest number at 845,767. Kathmandu also has the highest population density in the country at 5,108 persons per square kilometer. The rapid growth in population has led to mismanaged urbanization with little land use planning. At the same time, urban poverty is increasing due to poorly functioning land and housing markets, lack of planning for urban development and growth, and inflation.

Methods

This chapter is based on a literature review (both peer-reviewed and gray) on sources of lead, impacts of blood level on children and response measures for addressing blood lead level in Nepal. We used keywords such as lead contamination, children's health, Nepal, Blood Lead Level (BLL), policies and programs on Google Scholar database to identify relevant literature. We used articles published till 31 December 2022. We collected and systematized research on sources of lead, and patterns of exposure and explored underlying environmental injustices. We emphasize framing lead as an issue of child health disparity, social inequity, and environmental injustice.

Findings

Sources of Lead in Nepal

Until its ban in Nepal in the early 2000s, lead was commonly used in gasoline as an anti-knocking agent (Dhimal et al., 2017). There are now other sources of lead as it is widely used in electronics, ceramics, crystal glass, lead-acid batteries, cables, and paint.

Inadequate regulation and management of electronic waste, and substandard methods of recycling lead products (such as lead-acid batteries) put people at risk. One study shows that rag pickers working in Kathmandu who collect and process electronic waste are at increased risk of developing lead toxicity (Gautam et al., 2020).

According to the study of the Center for Public Health and Environmental Development (CEPHED) (2013), 54% of tested children's toys included toxic heavy metals, including lead, mercury, cadmium, chromium, and bromine. A sample of dust from school classrooms also contained more than 10 $\mu g/ft^2$ lead. This evidence points to the urgent necessity for continuous compliance monitoring, public education campaigns, and increased health sector participation.

One study on lead in Nepal showed how fruits grown in soil with high lead levels could pass this contamination on to consumers (Sah & Yadav, 2010). Soil contamination also comes from lead in pipes, lead dust from outdated paints, and leftover lead from gasoline.

The lead problem is particularly acute with lead in paint. For example, it was found that 71% of all paint products sold in Nepal had lead levels above 90 parts per million (ppm) (CEPHED, 2013). Another study found lead concentrations up to 200,000 ppm (20% lead by weight) in some brands of commonly available paints, indicating serious population-level exposure (Gottesfeld et al., 2014).

Lead paint is an important cause of lead exposure in children. As paints deteriorate, are subjected to friction, or are struck, they release tiny lead particles into the soil and in the air. Most lead exposure in children comes from lead in soil and dust pollution in homes with lead paint.

Schools and other spaces for children often use lead paint. One study examined 75 samples of paints in Nepal and 57 (76%) of those had lead contents greater than the national limit of 90 ppm. In addition, 53 (93%) of these paints had lead levels exceeding 600 ppm, and some of them had levels as high as 200,000 ppm (Gottesfeld et al., 2014).

A study from Kathmandu Valley shows that among the children with BLL greater than 5 μg/dl, 99.4% of them slept and played in a room with lead-based enamel paint (Dhimal et al., 2017). The study also found that 81.1% of children who had BLL ≥5 μg/dl, had their house painted with lead-containing paints (Dhimal et al., 2017). Furthermore, the same study reveals that 42.9% of children with elevated blood lead level (>BLL 5 g/dl) played with dirt and dust in or near their homes.

Enamel paint in particular is a problem. The maximum amount of lead found in enamel samples is almost 822 times greater than the US threshold for lead in paints.

The 13 enamel samples have an average lead concentration of 12,114 ppm which is approximately 135 times greater than the 90 ppm in Nepal's guideline for lead in paints (Sah & Yadav, 2010).

A study about the size of lead particles conducted along the Arniko Highway demonstrated that more than 90% of lead particles were larger than 5 μm, leading them to settle within 1.5 m of the road (and thus contaminating roadside houses and yards). Exposure from roads can range up to 1,000 ppm (Raj & Ram, 2013). Lower-income children in urban areas – especially those without backyards or nearby parks – are more likely to live and play by roadside that are lead dust contaminated.

These problems point to the need for policies to develop urban infrastructure and provide access to essential services including health care, schooling, sanitation, and a waste management system for the urban poor from environmental and health justice perspectives. Such policies could play a vital role in minimizing the exposure to lead as well as providing essential health care services for those affected.

Exposure to Lead and Blood Lead Level among Children in Nepal

Blood lead level is commonly used as an indicator for measuring exposure to lead in the environment. Table 23.1 presents key studies on exposure and blood lead level among children in Nepal.

Sherchand and colleagues (2014) revealed that among 218 school children from 6 to 16 years old in Kathmandu, 63% (137) had detectable BLL and 54% (117) had BLL >5 μg/dl. When assessing risk factors, the likelihood of having high BLL was significantly higher in children living in homes with peeling walls, children from families with lower socioeconomic status and who live near roads with heavy traffic (Sherchand et al., 2014).

Parajuli et al. (2014) found that lead exposure among children from birth to 36 months in Chitwan is high due to its location at the junction of the East-West Highway and two major routes into Kathmandu. They also found that BLL were not significantly associated with caste. Another study that did not find evidence of socioeconomic disparities was Parajuli et al. 2012. Conducted in the Terai with 100 mothers to evaluate the cord blood level of toxic and trace elements, the study used Brazelton Neonatal Behavioral Assessment Scale, Third Edition (NBAS III) and the Bayley Scales of Infant Development, Second Edition (BSID II) to evaluate the infant's neurological development at birth and at ages 6, 24, and 36 months. They classified caste based on surnames, which in Nepal is generally associated with one of the four main caste groups. The study found that cord blood levels of Lead, Zinc, Copper and Selenium were not associated with maternal age, socioeconomic status, living environment, and smoking status.

A descriptive cross-sectional study by Shrestha et al. (2021) conducted in Gokarneshwor Municipality in Kathmandu among 160 children between 8 and 18 years of age showed that children who frequently consume instant noodles, those who played with batteries, and those who had just undergone home renovations were associated with elevated BLL.

Table 23.1 Exposure and blood lead level among children in Nepal

Authors/Date	*Date*	*Title*	*Result*
Sherchand et al. (2014)	2014	Blood Lead Levels of Primary School Children in Kathmandu Municipality, Nepal	• 63% (137) had detectable BLL and 54% (117) had BLL $\geq$ 5 μg/dl. • Median BLL was 8 μg/dl in male (0–34) and 4 μg/dl in female (0–18).
Chandyo et al. (2017)	2017	Blood Lead Toxicity is not Associated with Anemia and Iron Deficiency in Children from Bhaktapur	• Mean blood lead concentration was 12.8 μg/dl. • 60% children had high lead concentration i.e., >10 μg/dl. • 63% of anemic children and 56% of non-anemic children had high blood lead concentration ($p = 0.2$). • 61% of iron deficient children and 54% of non-deficient children had high blood lead concentration ($p = 0.2$).
Parajuli et al. (2014)	2014	Impact of caste on the neurodevelopment of young children from birth to 36 months of age: a birth cohort study in Chitwan	• Caste was positively associated with the state regulation cluster score of NBAS III at birth after adjustment for covariates (p for trend < 0.01). • Adding cord blood as levels attenuated the association (p for trend = 0.12).
Shrestha et al. (2021)	2021	Blood Lead Level among Children between 8 and 18 years of Age by Atomic Absorption Spectrophotometry: A Descriptive Cross-sectional Study	• Mean blood lead level in children was 4.39 ± 7.35 μg/dl. • 18.75% (30/160) had blood lead levels.
Parajuli et al. (2015)	2015	Home environment and cord blood levels of lead, arsenic, and zinc on neurodevelopment of 24 months children living in Chitwan	• None of the BSID II cluster scores in 24-month-old infants were associated with cord blood levels of Pb. • The total HOME score was positively associated with the mental development scale (MDI) score (coefficient = 0.67, at 95% CI = 0.03 to 1.31). • A detrimental effect of in utero Pb on neuro-developmental indicators observed at birth disappeared at 24 months. • An association between neurodevelopment and home environment continued.
Parajuli et al. (2012)	2012	Cord blood levels of toxic and essential trace elements and their determinants in the Terai region of Nepal: a birth cohort study	• The mean values of Pb, in cord blood level were found as 31.7. • Cord blood levels of Pb, was not associated with maternal age, socioeconomic status, living environment, and smoking status.

(Continued)

Table 23.1 (Continued)

Authors/Date	*Date*	*Title*	*Result*
Mehta et al. (n.d.)	2015	Evaluation of lead toxicity among primary school children in Nepal	• 82% children (n = 503) had had detectable BLL and 64.7% (n = 396) had BLL > 5 µg/dl. • 7% children were at high risk of toxicity of lead exposure (>70 µg/dl). • Significantly negative correlation with blood hemoglobin (p = 0.002). • But only negative correlation with total calcium (−0.045) and IQ (−0.077). • Low socioeconomic status, increasing age, anemia, hypocalcemia and living in homes with chipped wall paints are potential factors associated with high BLL in children (p = 0.001) of Nepal.
Dhimal et al. (2017)	2017	High blood levels of lead in children aged 6–36 months in Kathmandu Valley, Nepal: A cross-sectional study of associated factors	• Of 312 children enrolled in the study, 64.4% had BLLs ≥ 5µg/dl. • A significant association was found between BLL and exposure to enamel paints in the household (p = 0.001). • Multivariate analyses showed that BLLs were 4.5 times higher in children playing with dirt and dust (p = 0.006). • Children belonging to the community of lower caste/ethnicity groups had significantly higher BLLs compared to those from the upper caste groups (p = 0.02).

Mehta et al. (n.d.) studied primary school children between 5 and 12 years living near high-traffic roads with air pollution (n = 612). The paper looked into children from ten schools of industrial region of eastern Nepal (n = 308) and others from eight schools of Kathmandu Metropolitan City (n = 304). Contrary to the two studies above (Parajuli, 2012, 2014), they found that low socioeconomic status, increasing age, anemia, hypocalcemia, and living in homes with chipped wall paints are potential factors that are significantly associated with high BLL in children of Nepal.

Likewise, a study by Dhimal et al. (2017) conducted among 312 children aged 6–36 months residing in the Kathmandu Valley found that more than 65% had high lead concentration in blood (BLL ≥ 5 µg/dl). 68.8% of the children having elevated blood lead level belonged to socially disadvantaged caste groups, compared to 31.2% who belonged to higher caste groups. Ethnicity was significantly associated with higher BLL in children. Furthermore, BLL were 4.5 times higher in children playing with dirt and dust and those belonging to lower caste/ethnicity groups had significantly higher blood lead level compared to those from the upper caste groups (Dhimal et al., 2017). But larger studies are needed to confirm the association between lead exposure and ethnicity in Nepal as most of the existing studies are small scale in nature and cover limited geographical areas. The findings from Dhimal et al. 2017 are bit different from other studies such as Sherchand et al., 2014, Meheta et al., n.d. which may be due to different study settings such as school children vs hospital enrolment and need further studies to confirm the role of caste/ethnicity and economic class on blood lead level among children in Nepal.

Many studies have also shown that elevated blood lead level negatively correlates with intelligent quotient (IQ). As such, lead toxicity can disrupt children's brain development (Dhimal et al., 2017; Mehta et al., n.d.; Parajuli et al., 2013; Sherchand et al., 2014). The parents of the children with elevated blood lead level ($>BLL \geq 5$ μg/dl) reported that their children were able to stand and walk later than expected while none of the parents of children with $BLL < 5$ μg/dl indicated developmental delays (Dhimal et al., 2017).

Environmental Health Injustice

In low-income counties, such as Nepal, lead is a serious public health issue. The limited regulation of polluting industries at the national and local levels, coupled with social hierarchies based on caste, ethnicity, gender, and class, all contribute to disparities in lead exposure.

Lead exposure among LMICs differs from those in high-income countries because of underlying social, cultural, economic, and historical factors (Kordas et al., 2018). Different studies show that different factors are at play in increasing lead exposure in Nepal, including socioeconomic status, environmental exposure, hazardous working environment, and cultural practices. Poor people are more likely to be exposed to lead because they either work in industries that expose them to lead or they live close to polluted sites. Limited rights, insufficient physical, social, or health protections, as well as a high level of work insecurity may exacerbate precarious environmental conditions associated with hazardous jobs. According to the risk-focusing model (Kordas et al., 2010), environmental exposures are "assigned" differently to groups with specific socially or culturally defined traits, such as people with low socioeconomic status and disadvantaged communities.

Education and employment are often the way out of poverty for ordinary families but it is particularly challenging for persons exposed to environmental toxins such as lead because their IQ levels are affected, among others. In other words, lead exposure continues the vicious cycle of poverty.

Lead exposure during childhood is associated with not only cognitive deficiencies but also a decline in social mobility over their lifetime. Consequently, it is important to emphasize the societal improvements required for upward mobility, especially given that conflicting national policies for economic growth, market pressures, and international competition may have dire consequences for poverty alleviation and hazardous work conditions. For example, disproportionate number of women and Dalits are involved in unskilled work that produces or uses harmful chemicals that can undermine household environment and health.

Response Measures

The World Health Organization (WHO) has urged the Government of Nepal to implement prompt, efficient steps to limit lead exposure, particularly among children in vulnerable populations.

Raising awareness can help. Since 2020, the International Lead Poisoning Prevention Week of Action is celebrated in October every year. This brings together government officials, paint manufacturers, environmental health campaigners, and child health advocates for collective action. Multiple stakeholders have also appealed to the Government of Nepal to adopt and effectively implement laws to protect children's health, working with the Global Alliance to Eliminate Lead Paint. In addition, the National Health Education, Information and Communication Center (NHEICC), the Ministry of Health and Population, the governments of Koshi, Madhesh, Bagmati, and Lumbini, as well as the WHO country office for Nepal, supported the Center for Public Health and Environmental Development (CEPHED) in Nepal to organize a number of awareness-raising and capacity-building programs at the provincial level under the theme of "Prevention of Lead Exposure through Effective Implementation of Lead Paint Standards in Nepal".

As a result of lead exposure on children and their widespread cognitive decline, a great deal of individual and national economic potential is lost, work productivity is reduced, and the likelihood of mental health issues rises. To this effect, governments in Nepal and elsewhere must act by passing strict regulation and policies when it comes to lead contamination. Effective regulation and policy can guarantee that the necessary workplace safety and industrial hygiene procedures are in place to help reduce or even eliminate lead exposure.

Lead can be recycled. With methods that adhere to the principles of the circular economy and closed-loop supply chains, it can be reused safely. Sites with lead contamination can be cleaned up and restored. Policymakers should prepare or reassess national action plans to address lead exposure based on sources of exposure identified with participation from relevant parties, including the paint industry. Spending more on healthcare workers' education to improve the diagnosis and treatment of lead exposure in children is essential. Children who experience developmental delays due to lead exposure should have access to a high-quality education and developmental therapies. The use of lead in different products such as paints, gasoline, and ceramics should be eliminated.

Vulnerable populations such as children and pregnant women should have legal protections to prevent them from being exposed to lead-containing products. The key to reducing long-terms effects of lead exposure is early intervention to reduce further exposure. Readily accessible primary medical care, frequent blood tests, and diets high in iron, calcium, and vitamin C can be effective to reduce the negative consequences of lead exposure. Private sector, governmental and non-governmental agencies, and other stakeholders should collaborate to protect the health of children, especially those in the most vulnerable social positions confronting gender, class, and caste discrimination.

Conclusion

Many studies in Nepal show elevated BLLs in children belonging to socially disadvantaged caste groups (Dalits, Janajatis and non-Dalit Terai caste groups, and relatively advantaged Janajatis too), those working in informal sectors, and poor children in occupations, such as rag picking. This elevates lead contamination from a health equity problem to an environmental justice crisis.

These disadvantaged groups face profound and intersecting socioeconomic challenges. The socioeconomic status of the family, in turn, can adversely affect the lives of children in many ways. In Nepal as well as in many other countries, poor people usually work in hazardous conditions. Children often accompany their parents to such hazardous areas where potential sources of lead may be present, and spend the day playing and sometimes working in that environment without any protective measures. Many of these children carry food to these places, without sanitary wrapping, and eat with their parents in the same environment without facilities to wash their hands. Together, these conditions put them at high risk of contamination: a clear case of environmental and social injustice. Strong enforcement of existing standards and the development of new higher health and occupational standards is needed to reverse this trend.

Acknowledgments

We are thankful to Mr. Prakash Dhimal for his contribution to literature search and for compiling findings of different related studies.

References

CEPHED (2013). *National Report: Lead in Nepal's New Enamel Household Paints.* Kathmandu: Center for Public Health and Environmental Development (CEPHED).

Chandyo, R. K., Shrestha, B., Zhao, B., Ulak, M., Shrestha, P. S., Shrestha, R., … & Shrestha, S. (2017). Blood Lead Toxicity Is Not Associated with Anemia and Iron Deficiency in Children from Bhaktapur, Nepal. *The FASEB Journal*, *31*, 786–2.

Dhimal, M., Karki, K. B., Aryal, K. K., Dhimal, B., Joshi, H. D., Puri, S., … & Kuch, U. (2017). High Blood Levels of Lead in Children Aged 6–36 months in Kathmandu Valley, Nepal: A Cross-Sectional Study of Associated Factors. *PLoS One*, *12*(6), e0179233.

Ericson, B., Hu, H., Nash, E., Ferraro, G., Sinitsky, J., & Taylor, M. P. (2021). Blood Lead Levels in Low-Income and Middle-Income Countries: A Systematic Review. *The Lancet Planetary Health*, *5*(3), e145–e153. https://doi.org/10.1016/S2542-5196(20)30278-3

Gautam, K., Pant, V., Pradhan, S., Pyakurel, D., Bhandari, B., & Shrestha, A. (2020). Blood Lead Levels in Rag-Pickers of Kathmandu and Its Association with Hematological and Biochemical Parameters. *EJIFCC*, *31*(2), 125.

Gottesfeld, P., Pokhrel, D., & Pokhrel, A. K. (2014). Lead in New Paints in Nepal. *Environmental Research*, *132*, 70–75.

Kordas, K., Queirolo, E. I., Ettinger, A. S., Wright, R. O., & Stoltzfus, R. J. (2010). Prevalence and Predictors of Exposure to Multiple Metals in Preschool Children from Montevideo, Uruguay. *Science of the Total Environment*, *408*, 4488–4494.

Kordas, K., Ravenscroft, J., Cao, Y., & McLean, E. V. (2018). Lead Exposure in Low and Middle-Income Countries: Perspectives and Lessons on Patterns, Injustices, Economics, and Politics. *International Journal of Environmental Research and Public Health, 15,* 2351. https://doi.org/10.3390/IJERPH15112351

Kosnett, M. J., Wedeen, R. P., Rothenberg, S. J., Hipkins, K. L., Materna, B. L., Schwartz, B. S., … Woolf, A. (2007). Recommendations for Medical Management of Adult Lead Exposure. *Environmental Health Perspectives, 115*(3), 463–471.

Mehta, K., Sherchand, O., Deo, B., Lamsal, M., & Baral, N. (n.d.). *Evaluation of Lead Toxicity among Primary School Children in Nepal.* Proceedings of the International Workshop on Molecular Biology Techniques Related to Infectious/Genetic Diseases & Human Identification, Molecular Medicine Unit, Faculty of Medicine, University of Kelaniya. http://repository.kln.ac.lk/handle/123456789/8786

Olufemi, A. C., Mji, A., & Mukhola, M. S. (2022). Potential Health Risks of Lead Exposure from Early Life through Later Life: Implications for Public Health Education. *International Journal of Environmental Research and Public Health, 19*(23). https://doi.org/10.3390/ijerph192316006

Parajuli, R. P., Fujiwara, T., Umezaki, M., Furusawa, H., Ser, P. H., & Watanabe, C. (2012). Cord Blood Levels of Toxic and Essential Trace Elements and their Determinants in the Terai Region of Nepal: A Birth Cohort Study. *Biological Trace Element Research, 147,* 75–83.

Parajuli, R. P., Fujiwara, T., Umezaki, M., & Watanabe, C. (2014). Impact of Caste on the Neurodevelopment of Young Children from Birth to 36 months of Age: A Birth Cohort Study in Chitwan Valley, Nepal. *BMC Pediatrics, 14,* 1–9.

Prevention of Lead Exposure through Effective Implementation of Lead Paint Standard in Nepal, ILPPWA 2020 Center for Public Health and Environmental Development. (n.d.). http://cephed.org.np/wp-content/uploads/2021/03/CEPHED-Report-of-ILPPW-2020.pdf

Raj, S. P., & Ram, P. A. (2013). Determination and Contamination Assessment of Pb, Cd, and Hg in Roadside Dust along Kathmandu-Bhaktapur Road Section of Arniko Highway, Nepal. *Research Journal of Chemical Sciences, 3*(9), 18–25.

Sah, R. C., & Yadav, K. (2010). *Study of Lead in Paints in Nepal.* Center for Public Health and Environmental Development (CEPHED).

Sherchand, O., Mehta, K. D., Poudel, P., Deo, B., & Baral, N. (2014). Blood Lead Levels of Primary School Children in Kathmandu Municipality, Nepal. *Journal of Institute of Medicine, 36*(3): 15–21.

Shrestha, Y. K., Khanal, M. P., Shrestha, S. K., Shrestha, J., & Shrestha, N. (2021). Blood Lead Level among Children between 8-18 years of Age by Atomic Absorption Spectrophotometry: A Descriptive Cross-sectional Study. *JNMA: Journal of the Nepal Medical Association, 59*(242), 1012.

Zamani, N., Hosseini, A., Farnaghi, F., Sayyari, A., Gholami, N., Imanzadeh, F., Hadeiy, S. K., Hajipour, M., Salimi, A., Philips, S., & Hassanian-Moghaddam, H. (2023). Blood Lead Level Evaluation in Children Presenting with Chronic Constipation in Tehran-Iran: A Cross-Sectional Study. *Scientific Reports, 13*(1), 1–7. https://doi.org/10.1038/s41598-023-29487-y

Index

Note: **Bold** page numbers refer to tables, *italic* page numbers refer to figures and page numbers followed by "n" refer to end notes.

For Product Safety Concerns and Information please contact our EU
representative GPSR@taylorandfrancis.com
Taylor & Francis Verlag GmbH, Kaufingerstraße 24, 80331 München, Germany

www.ingramcontent.com/pod-product-compliance
Ingram Content Group UK Ltd.
Pitfield, Milton Keynes, MK11 3LW, UK
UKHW022317250926
475702UK00046B/906

* 9 7 8 1 0 4 1 3 5 1 5 5 9 *